MANUAL T

Air Distribution Basics for Residential and Small Commercial Buildings

Air Conditioning Contractors of America
1330 Braddock Place, Suite 350
Alexandria, VA 22314
PHONE: 703 • 575 • 4477
FAX: 703 • 575 • 4449
www.acca.org

Reprinted 2021

ACKNOWLEDGMENTS

The purpose of this manual is to provide a comprehensive guide to sizing and selecting residential and small commercial supply air outlets and returns.

The development and publication of this manual was a joint effort by the **NORTHAMERICAN HEATING AND AIR-CONDITIONING WHOLESALERS ASSOCIATION (NHAW)** and the **Air conditioning Contractors of America (ACCA).** Much of the information in this manual is secured from, or based on information published in the **American Society of Heating, Refrigeration and Air-Conditioning Engineers (ASHRAE)** handbooks and from various design manuals and engineering data manuals published by equipment and air terminal manufacturers.

Grateful acknowledgment is made to those members of both the **NHAW** and **ACCA** who contributed to the preparation of this manual. Gratitude is also extended to **Hart & Cooley, Inc,** and **Titus Products** for providing the product performance data that was used.

This manual is based on a home study course that is offered by the **NHAW**. The original manuscript for the home study course was written by **James H. Healy** of the **NHAW.**

The home study course manual was converted into an **ACCA** manual by **Hank Rutkowski, P.E., ACCA Technical Director.**

Review and Technical Assistance:

James H. Healy
Northamerican Heating and Air-Conditioning Wholesalers Association
Columbus, Ohio

Bill Wright
Lexington, Massachusetts

ACCA Manual T Task Committee
Clint Van Der Kolk
Chairman, ACCA Technical Services Task Team
Kalamazoo, Michigan

Lynn Cawthorne
Nashville, Tennessee

Warren Lupson
Forestville, Maryland

Fred Piff
Mobile, Alabama

Vern Ramsey
Indianapolis, Indiana

Jon Turner
Winston-Salem, North Carolina

ISBN: 1-892765-06-3

Introduction

The selection and sizing of supply air outlets and return air inlets is just as important to a successful design as making load calculations, selecting equipment, or sizing ducts. Supply air outlets must mix the raw supply air with the room air without creating drafts or stagnate zones. Return air inlets must provide an unrestricted return air path from every area of the house or building without short circuiting the supply air pattern.

In addition, properly located supply air outlets can help offset cold drafts that are normally associated with windows and glass doors or the ankle-level discomfort that is often experienced with slab-floor construction.

This manual provides the basic information that is required to correctly select, size and locate constant volume supply out-lets and return inlets. Floor, low-side wall, high-side wall and ceiling outlets are discussed. The procedures outlined in this manual apply to either residential or commercial work.

This manual **(Manual T)** replaces all current and previous editions of the **ACCA Manual E, "Room Air Distribution Considerations." Manual T** includes all of the information that was provided in **Manual E** plus a considerable amount of new material. This new material extends the discussion of principles and theory that was initiated in **Manual E.** But more important, **Manual T** includes new material which documents application guidelines and selection procedures. These guidelines and procedures are explained by using manufacturer's data to solve a number of example problems.

How to Use this Manual

This manual is intended to provide a primer on the various factors that affect room air distribution. The principles and procedures outlined apply to the basic types of constant volume supply air outlets and return air inlets that are commonly installed as part of a residential or a small commercial air distribution system. These basic types include floor, low-side wall, high-side wall and ceiling outlets and/or inlets.

Manual T does not discuss the application of various types of air-side devices that are often found on large commercial, industrial or institutional air distribution systems. A partial list of such devices includes long slot diffusers, perforated ceiling panels, VAV terminals, light troffers, VAV boxes, constant volume/VAV fan boxes, constant volume and variable volume reheat boxes, constant volume and variable volume mixing boxes, system powered air side devices, ceiling mounted induction boxes, air valves, and air flow or pressure measurement stations.

Everyone who uses this manual should study Sections 10 and 11. These sections summarize the methods and procedures that are used to select and size outlets and inlets. The remaining sections of this manual provide more detail on and explanation of the various factors that affect outlet and inlet performance.

Also note that Appendix 2 provides a summary of the various equations charts and tables that are used to select and size outlets and inlets.

Table of Contents

Acknowledgements i
Introduction iii
How to Use This Manual iv

Section 1

What Makes People Comfortable 1-1

Minimum Amount of Air Required for Survival 1-1
Amount of Air Required for Comfort 1-1
Metabolic Rate 1-1
Heat Transfer to the Surrounding 1-1
Effect of Surrounding Temperature 1-2
Building Envelope Affects Human Comfort 1-2
Mean Radiant Temperature 1-2
Effect of Winter Humidification 1-3
Effect of Velocity 1-3
Natural Convection 1-3
ASHRAE Comfort Chart 1-4
Temperature Gradients 1-5

Section 2

Air Distribution Concepts and Terminology 2-1

General Categories of Terminal Devices 2-1
Types of Ceiling Diffusers 2-1
Types of Floor Diffusers 2-2
Primary Air and Secondary Air 2-2
Throw 2-2
Drop 2-3
Rise 2-3
Surface Effect 2-3
Spread 2-3
Primary Air Pattern 2-4
Total Air Pattern 2-4
Entrained Air 2-4
Stagnate (or Stratified) Zone 2-4
Vertical Temperature Gradient 2-5
Gross Area 2-5
Free Area 2-5
Face Velocity 2-6
Core Velocity 2-6
Velocity/Flow-Rate Equation 2-6

Section 3

High Sidewall Terminals 3-1

Location 3-1
Primary Air Pattern 3-1
Entrance Condition Affects the Air Pattern 3-1
Total Air Pattern 3-2
Stratification 3-2
Secondary Air Pattern 3-2
Use Vanes to Control Air Pattern 3-3
Characteristics of High Sidewall Registers 3-3
Location of Returns 3-4

Section 4

Ceiling Diffusers 4-1

Primary Air Pattern, Radial Discharge 4-1
Primary Air Pattern, Directional Discharge 4-1
Primary Air Pattern, Vertical Discharge 4-2
Entrance Conditions Affect the Air Pattern 4-2
Total Air Pattern, Horizontal Discharge 4-2
Total Air Pattern, Vertical Discharge 4-3
Stratification, Horizontal Discharge 4-4
Stratification, (Vertical Discharge Diffusers) 4-4
Characteristics of Horizontal Discharge Outlets 4-4
Characteristics of Vertical Discharge Outlets 4-5
Location of Returns 4-6

Section 5

Floor and Low Sidewall Outlets 5-1

Primary Air Pattern, Vertical Diffuser 5-1
Primary Air Pattern, Vertical Registers 5-1
Primary Air Pattern, for Horizontal Discharge 5-2
Entrance Conditions Affect the Air Pattern 5-2
Total Air Pattern, (Vertical Discharge Diffusers) 5-2
Total Air Pattern, (Vertical Discharge Registers) 5-3
Total Air Pattern, (Horizontal Discharge Outlets) 5-3
Basement Heating 5-4
Stratification, (Vertical Discharge Diffusers) 5-4
Stratification, (Vertical Discharge Registers) 5-5
Stratification, (Horizontal Discharge Outlets) 5-5
Floor and Low Sidewall Outlet Characteristics 5-5
Location of Returns 5-7

Section 6

Pressure Loss Characteristics of Supply Outlets 6-1

Inches of Water 6-1
Static Pressure 6-1
Total Pressure 6-1
Velocity Pressure 6-1
Static Regain 6-2
Pressure Loss with Constant Velocity 6-2
Pressure Loss with Changing Velocity 6-2
Pressure Loss of Supply Outlets 6-2
Accessory Pressure Loss 6-3
Optimum Outlet-Fitting Geometry 6-3

Section 7

Performance of Return Inlets 7-1

Types of Return Inlets 7-1
Negligible Effect on Room Air Patterns 7-1
Guidelines for Locating Returns 7-2

Section 8

Pressure Loss Characteristics of Return Inlets 8-1

Inlet Pressure Loss 8-1
Pressure Loss Is Affected by Inlet Geometry 8-1

Section 9

Sound Considerations **9-1**

Frequency 9-1
Sound Pressure 9-1
Sound Power 9-1
Frequency Spectrum 9-1
Loudness Chart 9-2
Noise Criteria Rating 9-2
Published NC Ratings 9-2
Acceptable NC for Various Applications 9-3
Face Velocity Guidelines 9-3
Catalog Velocity Recommendations 9-4
Installation Details Affect Noise Levels 9-4
Integral Dampers Generate Noise 9-4
Attenuation 9-5
Generation 9-5
Equipment Near the Occupied Space 9-5

Section 10

Selecting Supply Outlets **10-1**

Objectives 10-1
Location 10-1
Approximate Air Terminal CFM 10-1
Selection Procedure 10-1
Affect of Fan Operation 10-2
Noise Criteria 10-2
Manufacturer's Data 10-2
Floor, Baseboard and Low Sidewall Outlets 10-3
Example: Floor or Baseboard Outlet Selection 10-3
Ceiling Diffusers 10-4
Example: Nondirectional Ceiling Diffuser 10-5
Example: Directional Ceiling Diffuser 10-6
High Sidewall Outlets 10-6
Example: High Sidewall Outlet Selection 10-8

Section 11

Selecting Return Inlets **11-1**

Style and Location 11-1
Size 11-1
Pressure Drop 11-1
Transfer Grilles 11-1
Filter Grilles 11-2
Residential Example 11-2
Commercial Example 11-2
Transfer Grille Example 11-2

Section 12

Balancing and Measuring Air Flow **12-1**

Balancing procedures 12-1
Area Factor 12-1

Appendix 1

Manufacturer's Application Data **A1-1**

Appendix 2

Summary of Tables, Charts and Equations **A2-1**

Section 1

What Makes People Comfortable

Proper air terminal sizing and selection procedure is closely related to the various factors that affect human comfort. Therefore, it is appropriate to begin the discussion with a review of what makes people comfortable.

1-1 Minimum Amount of Air Required for Survival

Humans require surprising little air to sustain life. During normal breathing one cubic foot of air per minute is quite adequate to replenish the oxygen consumed by a person and to prevent objectionable concentrations of exhaled carbon dioxide from building up.

1-2 Amount of Air Required for Comfort

While people need very little air to survive, they need considerable quantities of properly conditioned air to be comfortable. A great many factors, both physical and psychological, are associated with being or feeling comfortable. The heating and air conditioning practitioner must be concerned with the factors affecting the conditions which provide a healthy and thermally comfortable environment for people.

1-3 Metabolic Rate

The energy (calories) contained in the food we eat is released when the food is oxidized or in a sense burned inside the body. This process of internal heat generation is called metabolism. This internally generated heat must be dissipated, otherwise body temperature would rise and a person would become ill or die. Some of this internal energy is released as work (moving about) but most of the energy is converted to heat which is transferred to the surroundings. Of course, the exact amount of heat generated by a person depends on the level of activity. The figure below provides a few examples of how the heat that is generated by a person is related to their activity level.

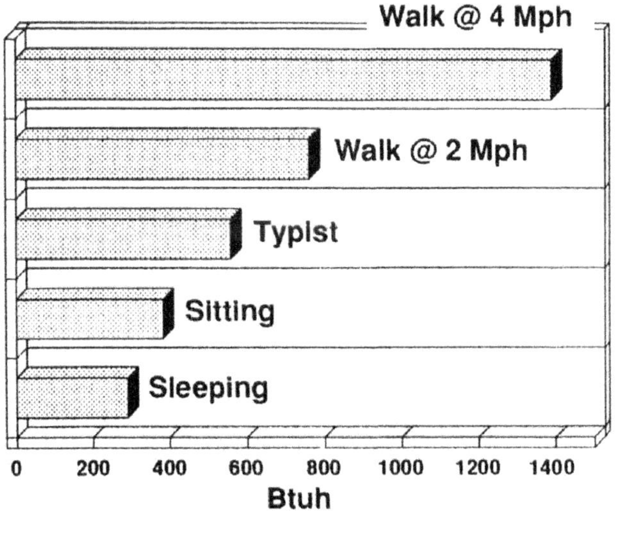

1-4 Heat Transfer to the Surroundings

The deep tissue within the human body operates at approximately 98.6 degrees and under normal conditions, the body's "cooling system" controls the body temperature so that it is always within a few tenths of a degree of that temperature. (However, the skin surface temperature is typically cooler, approximately 86 degrees.) Thus the first rule for keeping a person comfortable is to maintain the environment so that the heat that is generated by the metabolism can be rejected at a "comfortable" rate. Allow people to lose heat too fast and they shiver and shake; too slow and they get warm and perspire. The body's control system reduces heat transfer to the surroundings when the body temperature begins to drop and increases the heat transfer when the body temperature rises. Heat is transferred from the body to its surroundings by convection, evaporation, radiation and (to a small extent) conduction.

- Heat Loss by Convection

As the temperature of the human body rises from its normal 98.6 F, the blood vessels near the surface of the skin expand allowing more blood to flow to the skin. The skin temperature rises due to the presence of the hotter blood and this increases the quantity of heat lost to the air because of the greater skin-to-air temperature difference. The skin serves as the body's "radiator." The temperature and velocity of the air moving over the skin has a direct affect on the amount of convective heat loss from the skin.

- Heat Loss by Evaporation

Should the heat lost by convection be inadequate to remove all of the excess heat from the body, the deep body temperature will continue to rise causing the sweat glands to open. As the surface of the skin becomes moist, additional heat is removed as the sweat is evaporated into the surrounding air. In the process, the skin surfaces are cooled. The amount of evaporation (cooling affect) from the skin depends on the humidity and velocity of the surrounding air. The higher the humidity, the lower the rate of evaporation. Increasing air motion tends to increase the rate of evaporation.

- Heat Loss by Radiation

Radiation is perhaps the least understood method of body heat rejection but it is a very important mode of heat loss. The classic example of radiation heat transfer is the heat received from a glowing fireplace. In this process, your body is warmed directly by the radiant energy emitted by the fire and not by the air that is in front of you. For radiant heat transfer to occur, the body must "see" the radiating surface. The human body receives heat energy from any surface that is warmer then the skin and loses heat to any surface that is cooler than the skin.

For example, in the winter, the inside surface of a large window may be much colder than a person's skin surface temperature, hence the body would radiate considerable heat to the cold window. A person may even shiver in front of the large window despite the fact that the room air temperature is normal and body heat loss by convection and evaporation is minimal. To compensate for the high radiation heat loss, a person might put on a sweater or raise the thermostat to increase room air temperature and further reduce convective and evaporative heat loss.

1-5 Effect of Surrounding Temperature

The curves below illustrate how the amount of heat rejection by each heat transfer mode can change depending on the surrounding environment. Note as the air temperature increases, body heat loss by evaporation increases and heat loss by radiation and convection diminishes.

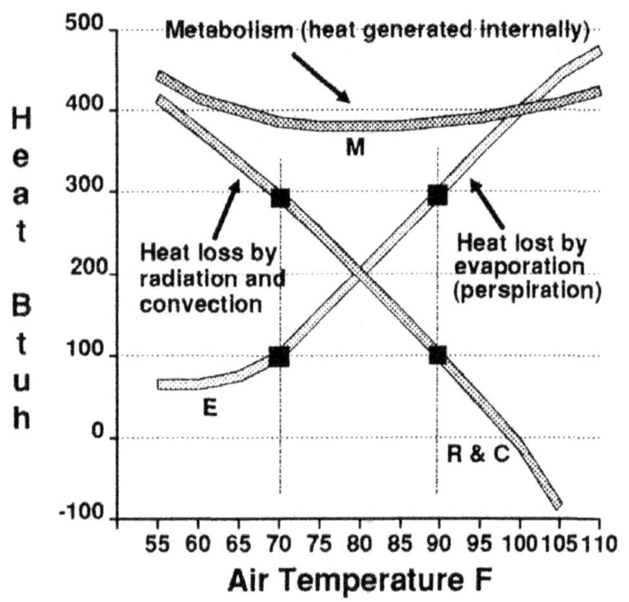

At 70 F, only 100 Btu/hr is lost by perspiration and 300 is lost by radiation and convection. At 90 F, the situation is reversed, less heat is carried away by radiation and convection and more heat is carried away by evaporation.

1-6 Building Envelope Affects Human Comfort

The quality of the building/room thermal envelope is obviously a major factor in providing human comfort. Insulation, for example, can reduce radiant heat loss by increasing the inside wall surface temperatures.

The illustration below shows the temperature gradient through a typical frame wall both insulated and uninsulated. On very cold days, the inside uninsulated wall surface temperature might be as low as 61 degrees while a fully insulated wall at the same outdoor conditions would be over 71 degrees. This means that a person would radiate less heat to an insulated wall than an uninsulated wall. It also means that designing an air distribution system which will provide a comfortable environment would be an easier task.

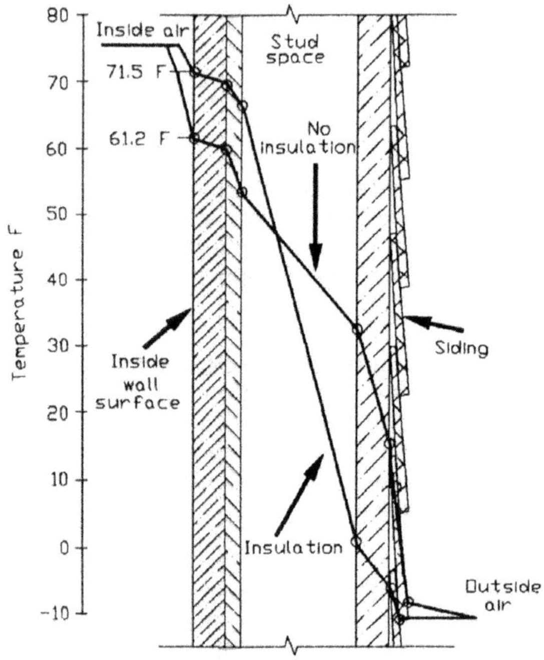

For example, a well insulated envelope would provide a winter environment that consists of 75 F room air with inside surface of the walls at a little below 70 F. The illustration below shows how a person seated at rest might lose heat to his environment. About 200 Btu/hr would be lost by radiation to the colder walls and convection to the room air; perhaps 100 Btu/hr would be lost through evaporation. If a large amount of glass was installed or if the walls were uninsulated the wall surface temperatures would be less than 70 F and the person would lose more heat by radiation.

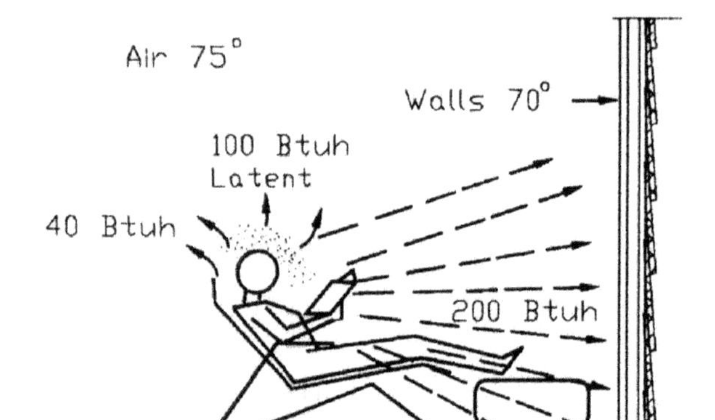

1-7 Mean Radiant Temperature

Laboratory tests show that surface temperatures are as important as room air temperature to overall comfort. In fact researchers suggest that increasing or decreasing surface temperatures 1 degree is the same as increasing or decreasing room air temperature 1 degree in terms of the comfort achieved.

The mean radiant temperature (MRT) is one way to measure the effect of the inside surface temperatures. The MRT temperature can be derived from equations which consider the various room surface temperatures or by using a device called a "globe" thermometer which is illustrated in the figure below.

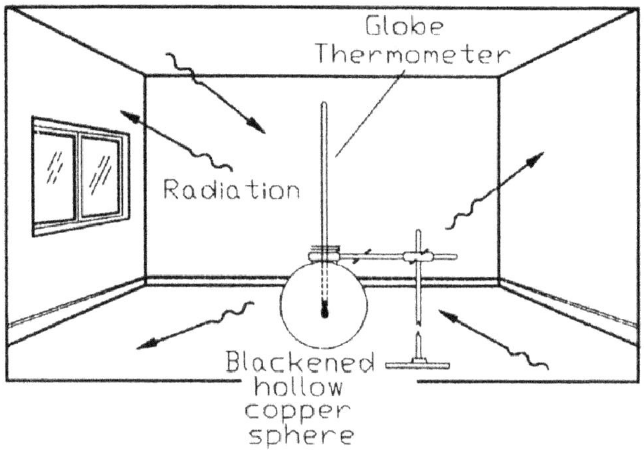

Experts in measuring human comfort often combine the room air temperature with a room's mean radiant temperature to create an improved index of comfort. This new index is called the "adjusted dry-bulb" temperature. The adjusted dry-bulb temperature is calculated by averaging the room air temperature and mean radiant temperature.

For example, the adjusted dry-bulb temperature for a room at 75 F with an MRT of 71 would be:

$$(75 + 71) / 2 = 73 F$$

The fact that the thermal characteristics of the building envelope is so important in maintaining comfort means that the designer must consider inside surface temperatures (especially the amount, type and location of glass or poorly insulated walls) when the air distribution system is designed. This will be discussed in more detail later.

1-8 Effect of Winter Humidification

As noted in section 1-4 above, the indoor humidity does affect the rate of body heat rejection by evaporation. However, If only residential and office environments are considered, the effect of large changes in indoor humidity are minimal. During the heating season, a change from 20% to 60% RH may go practically unnoticed by most occupants. The important point is that while humidity is a factor in heating season comfort, the human body is more sensitive to slight changes in air and surface temperatures than it is to changes in humidity. (Of course during the cooling season people have a low tolerance for high humidity and humidity control becomes more critical.)

1-9 Effect of Velocity

The human body is very sensitive to air motion, especially when seated at rest, say watching TV, reading, working at a desk, etc. A draft is defined as air with a specific combination of air velocity and air temperature that causes a sensation of coolness, hence discomfort.

How fast must air move to cause a draft? It depends on which part of the body is affected by the moving air stream. The chart below outlines comfort zones for the neck and ankles for a range of air temperatures and air velocities. At 72 F, air moving at 30 feet per minute (fpm) would be outside the comfort zone near the neck and constitute a draft. That same 72 F air around the ankles could be moving at 90 fpm or more and would not be felt as a draft. As air gets cooler, velocities must be reduced to avoid discomfort.

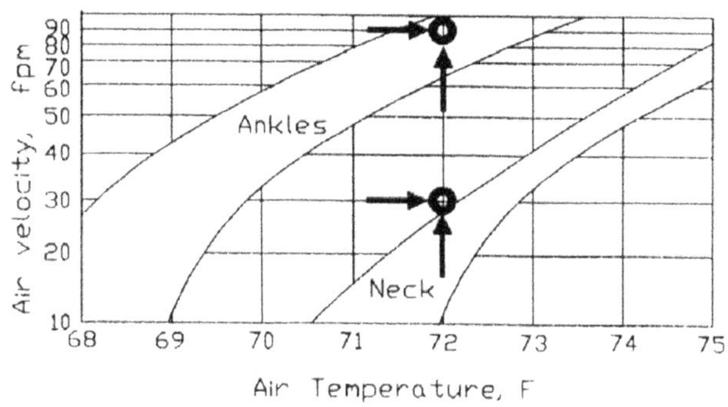

The following chart provides some approximate relationships between air velocities and physical experiences. As a further comparison, the velocity of the air discharging from a floor register might range from 300 to 500 fpm.

Velocity	Result
15 fpm	Cigarette smoke hangs in air
65 fpm	Paper might blow off of desk
160 fpm	Hair, light objects blow about

1-10 Natural Convection

Usually there is air motion in a room as a result of "natural" conditions. In winter, room air contacts cold glass and cold exterior wall surfaces. This cools the air. Since the cooler air is heavier it moves down the wall and across the floor.

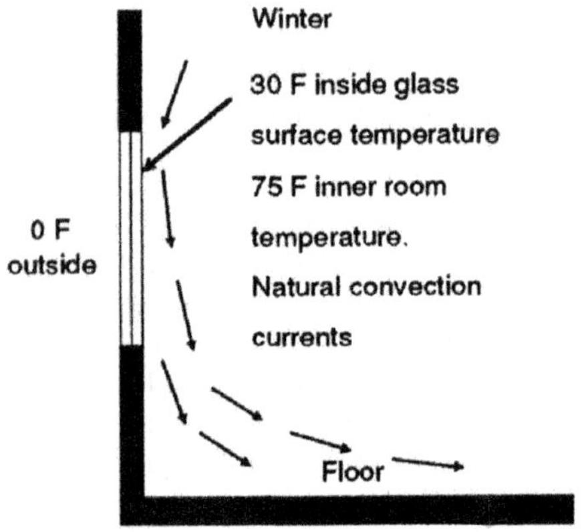

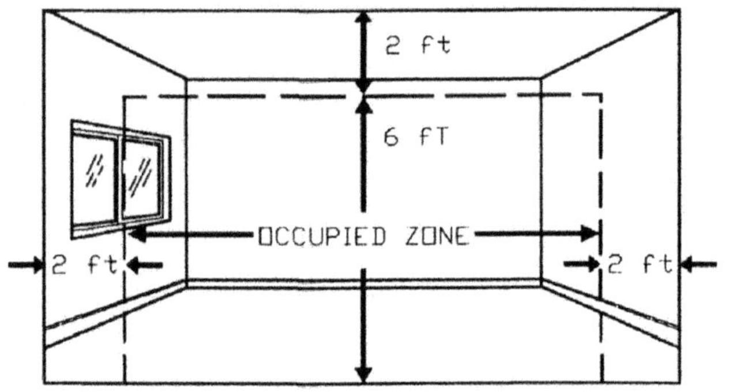

In summer, the reverse occurs. Room air in contact with the warmer glass and walls is heated and becomes lighter. This air moves upward and across the ceiling. Lights and appliances can produce their own local air currents as well. Such air motion is referred to as natural convection (air) currents.

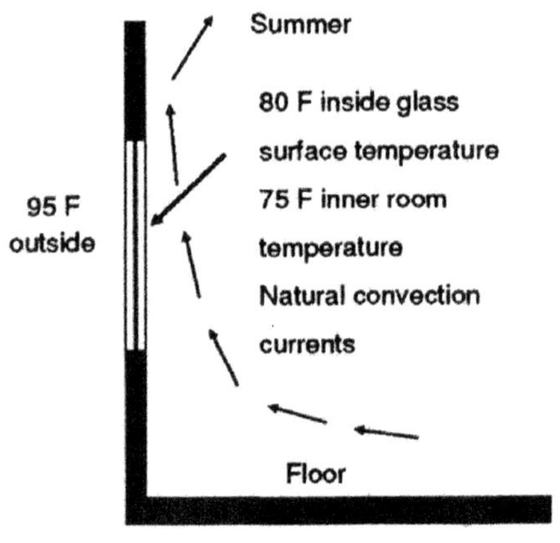

1-11 ASHRAE Comfort Chart

The American Society of Heating, Refrigerating, and Air-Conditioning Engineers' (ASHRAE) Standard 55-1981 specifies the conditions under which 80 percent or more of the occupants would be comfortable. The standard points out that clothing is an important modifier of body heat loss and comfort. Because people dress differently in winter and summer, the indoor conditions under which people feel comfortable differ slightly between summer and winter.

An important part of the ASHRAE comfort standard is the concept of the "occupied zone." The occupied zone consists of the space between the floor and 6 feet above the floor in the vertical direction and the space that is more than 2 feet from the wall in the horizontal direction. It is within the occupied zone that comfort conditions must be maintained.

The figure below is a reproduction of the ASHRAE comfort chart. This chart summarizes the conditions which define comfort during the heating season and during the cooling season.

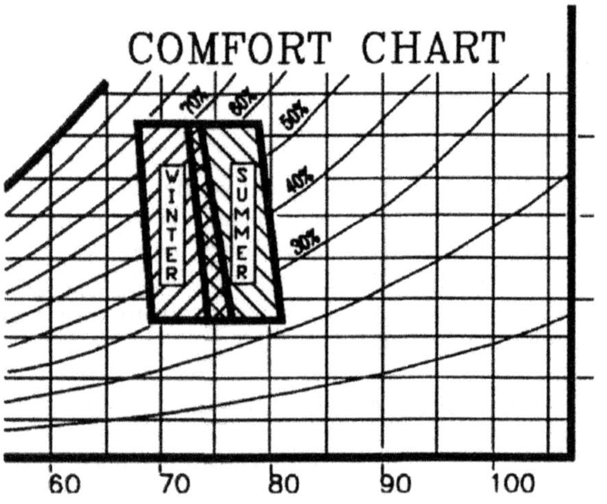

In the winter, the adjusted dry-bulb temperature which you recall is the average of the air temperature and the mean radiant temperature, should be kept between 67 F and 73 F. In a poorly insulated building with resulting cold wall surface temperatures, the air temperature would have to be raised to maintain comfort. Air motion should average 30 fpm.

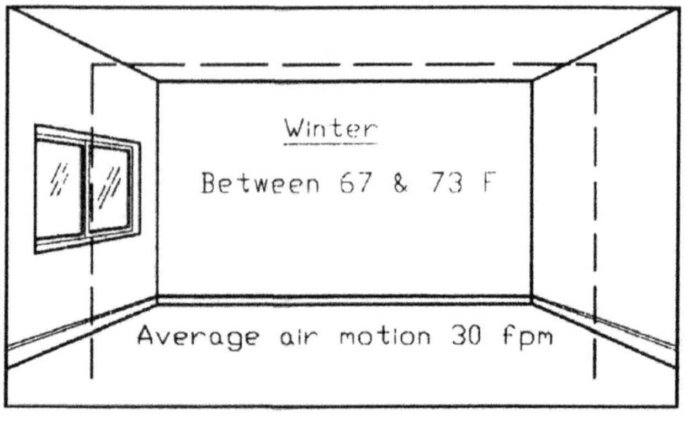

In the summer, the adjusted dry-bulb temperature should range from 73 F to 79 F with the room air motion slightly higher than in winter, averaging 50 fpm. (In some nonresidential applications, the adjusted dry-bulb temperature can be increased 1 F for every 30 fpm increase in average air motion up to 160 fpm. Refer to **Industrial Ventilation**, 13th edition or later.

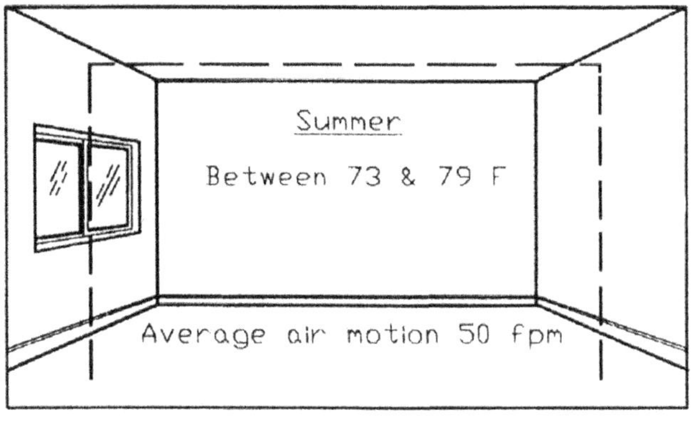

1-12 Temperature Gradients

Air temperatures in a room generally increase from floor to ceiling. To prevent local discomfort, the vertical temperature difference between the 4-inch level and the 67-inch level should not exceed 5 F. For example, if the temperature at the 67-inch level was 73 F, then the temperature at the 4-inch level should not be less than 68 F.

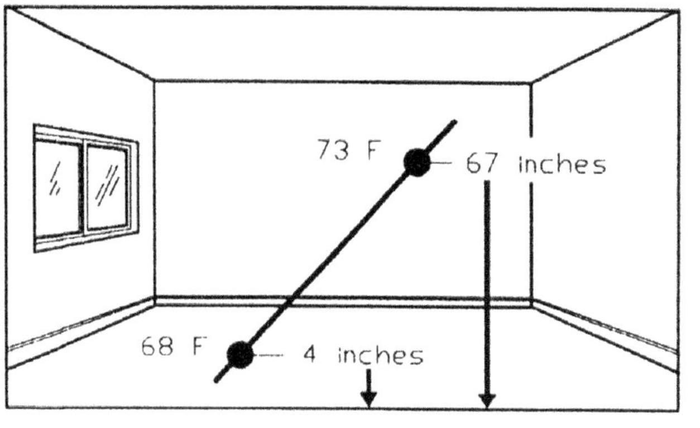

Section 2

Air Distribution Concepts and Terminology

Central heating and cooling equipment is connected to a network of ducts which distribute the supply air to each room. Each supply duct terminates at a supply outlet which mixes the supply air with the room air before the supply air enters the occupied zone. The supply outlets perform an important but difficult task because the supply air temperature is usually much hotter or colder than room temperature, and because the discharge air velocity is about 10 times greater than the velocity that is desired within the occupied zone.

The central equipment is also connected to one or more return air ducts. Each return run begins at a return intake. The job of the return air intake is to provide an unrestricted path for the air back to the conditioning equipment for reprocessing and recirculation.

The size, type and placement of a supply outlet has far greater importance on the conditions in the occupied zone (comfort) than the return intake does. Careful selection, sizing and placement of the supply air terminals is absolutely necessary. Sizing the return inlets is also very important. Return inlet placement is less critical but still requires consideration.

2-1 General Categories of Terminal Devices

Supply and return terminals fall into one of three general categories which are "grilles," "registers" and "diffuser."

- A grille is any louvered or perforated covering which fits over the open end of a supply duct, return duct, or over an opening that has been cut out of a plenum. Supply grilles are usually located in a wall or floor and normally are not installed in the ceiling. Return grilles can be located in a wall, floor or ceiling. Louvered grilles are normally fitted with horizontal and/or vertical vanes which may either be fixed in one position or adjustable. Grilles are almost always square or rectangular in shape. Supply grilles are usually located in a wall or floor. Return grilles can be located in a wall, floor or ceiling.

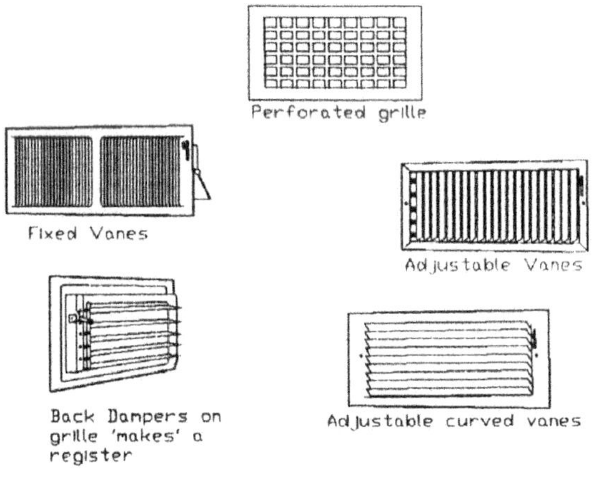

- A register is simply a grille fitted with some kind of air volume control (damper) which can be used to regulate the amount of air flowing through the device. Registers can be used as supply outlets or return inlets. (Using a register as a return inlet provides a means for making minor adjustments in the return air flow.) Supply registers are usually located in a wall or floor. Return registers can be located in a wall, floor or ceiling.

- A diffuser is a specialized supply outlet usually located in the ceiling but which may occasionally be located in the floor or side wall. Diffusers can handle relatively high air flow rates without creating drafts because they have more "mixing" ability than grilles and registers. Diffusers may or may not be equipped with a damper and can have a variety of shapes ranging from round to long and narrow. A diffuser could be used as a return inlet but a grille or register would be more appropriate.

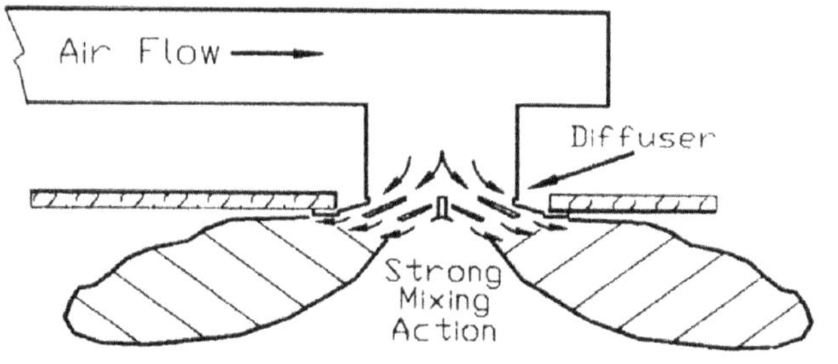

2-2 Types of Ceiling Diffusers

Ceiling diffusers are available in models which are designed to either discharge the air horizontally, parallel to the ceiling, or vertically toward the floor. The horizontal discharge models may either be designed to discharge the air in a non-directional, circular pattern or to discharge the air in a one-way, two-way, three-way or four-way directional pattern. Horizontal discharge diffusers may be round, square, rectangular or long and narrow. Long narrow ceiling diffusers can be further classified as linear diffusers or slot diffusers. Some types of horizontal discharge ceiling diffusers are designed to "lay in" between the T-bar openings of a suspended ceiling. Lay-in diffusers often feature perforated faces and adjustable pattern controllers which provide a means to control the direction of the discharge air. The vertical discharge models may be round, square, rectangular or a linear type or a slot type.

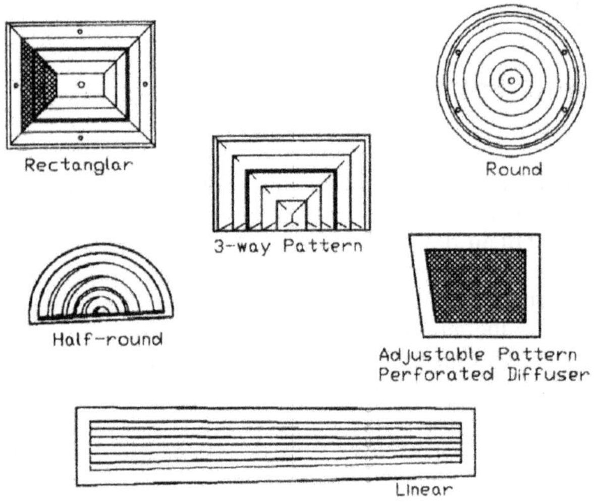

2-3 Types of Floor Diffusers

Floor, baseboard and low side-wall diffusers were developed to introduce draft-free air vertically, parallel to the wall. These devices are usually used for residential heating and cooling.

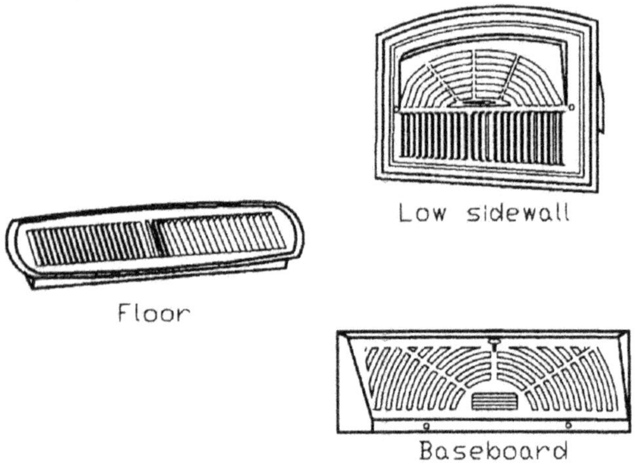

2-4 Primary Air and Secondary Air

When a duct without a supply terminal (inserted in the middle of a wall not adjacent to a ceiling or a floor) delivers air at exactly room temperature, a cone-shaped pattern of moving air will develop. This pattern expands in girth as the supply air (primary air) induces or draws in room air (secondary air) into the pattern. Depending upon the distance from the duct opening the total volume of air in motion (the sum of the primary and secondary air) may be 10 to 20 times greater than the volume of primary air. Thus, if the duct were delivering 80 cubic feet of air per minute (CFM) of primary air, at 12 feet away, 800 to 1600 CFM would be set in motion.

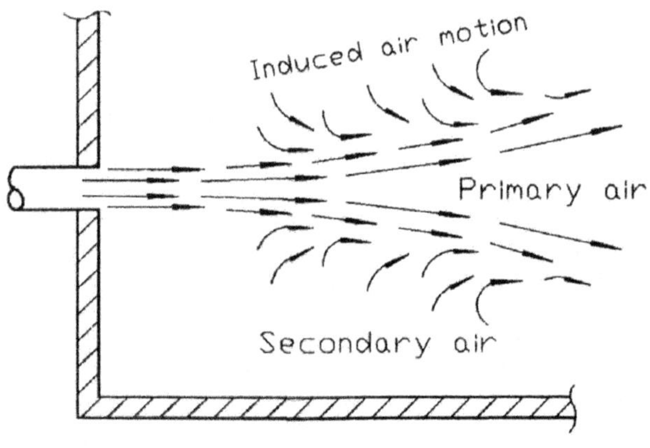

2-5 Throw

An important characteristic of a moving air stream is called throw. Throw is the horizontal distance (vertical distance in the case of a vertical discharge terminal), an air stream must travel after leaving the terminal until its velocity reaches some specific value, say 50 fpm.

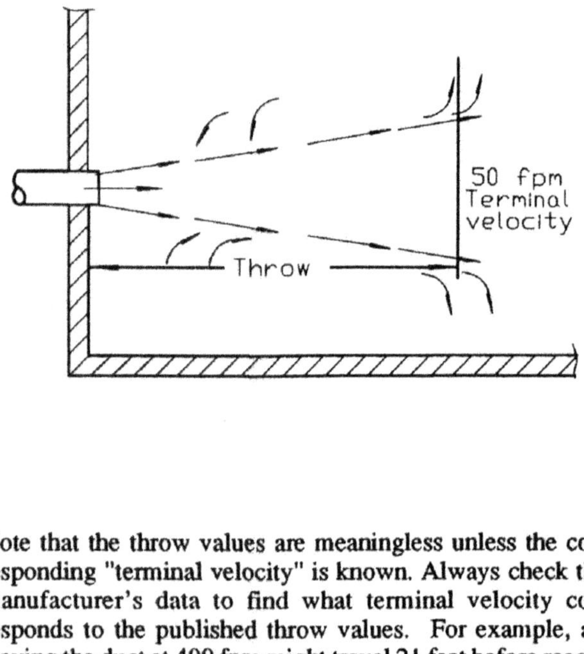

Note that the throw values are meaningless unless the corresponding "terminal velocity" is known. Always check the manufacturer's data to find what terminal velocity corresponds to the published throw values. For example, air leaving the duct at 400 fpm might travel 21 feet before reaching a velocity of 50 fpm. The throw would then be 21 feet (at a terminal velocity of 50 fpm). The same air stream might only travel 15 feet before reaching a velocity of 100 fpm. In this case the throw would then be 15 feet (at a terminal velocity of 100 fpm).

When ceiling terminals discharge the air in a radial direction the throw may be referred to as the "radius of diffusion" or "radius." In many cases the manufacturer may provide information on the minimum radius (higher value of terminal velocity) and the maximum radius (lower value of terminal velocity).

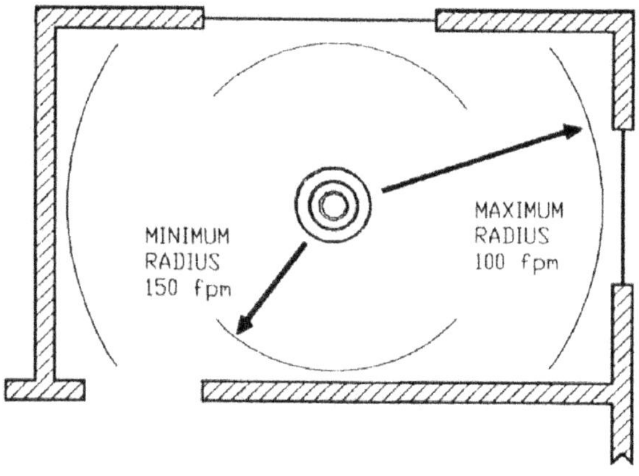

At any given CFM, the throw is basically a characteristic of the size, design and in some cases, the vane setting of the supply terminal. However, the temperature difference between the room air and the supply air also affects the throw, as does a parallel wall or ceiling surface that is located close to the supply outlet. All of this will be discussed in more detail below.

2-6 Drop

When the primary air is colder than the room air, the air pattern tends to drop down into the room. Drop is an important characteristic which is defined as the distance the bottom edge of a horizontally projected air stream falls by the time the air reaches the end of its throw. The tendency for the air pattern to drop increases as the temperature difference between the room air and the colder supply air increases. Also note that the drop that is associated with cooler primary air reduces the throw when compared to the throw that would occur if the primary air were at a neutral temperature.

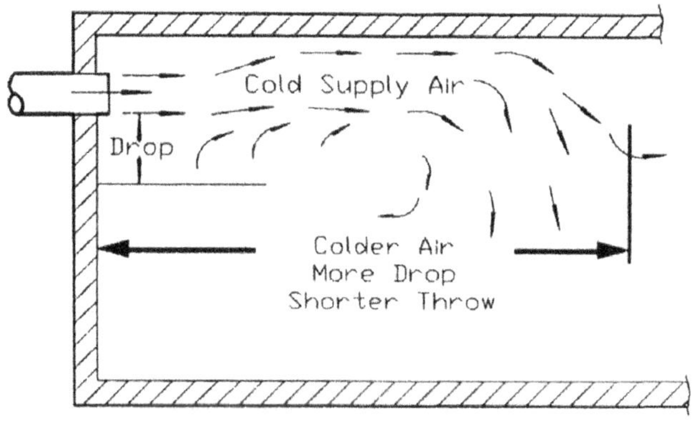

2-7 Rise

When a horizontally projected primary air stream is warmer than the room air, the air stream tends to rise above the level of the supply outlet. This rise also affects the throw. In this case the throw tends to decrease as the difference between the warm supply air and the room air increases.

2-8 Surface Effect

As noted in the section above, unobstructed jet stream expands in all directions at an angle of 22 degrees (Figure A below). However if the duct is relocated near the ceiling, the jet cannot now expand in all directions and on the ceiling side of the air stream, the entrainment of room air is restricted (Figure B below). This creates a low pressure region between the air stream and the ceiling. This low pressure "pulls" the air stream to the ceiling. The net effect is to reduce the amount of entrained air and to lengthen the throw as compared to an unobstructed air stream. As long as the duct opening is within one foot down from the ceiling away, this so- called "surface effect" will occur. The surface effect also occurs when a floor or ceiling opening projects air parallel to a wall. As we'll learn later, this effect will enhance the ability of perimeter heating and cooling systems to supply air to the room without creating drafts within the occupied zone.

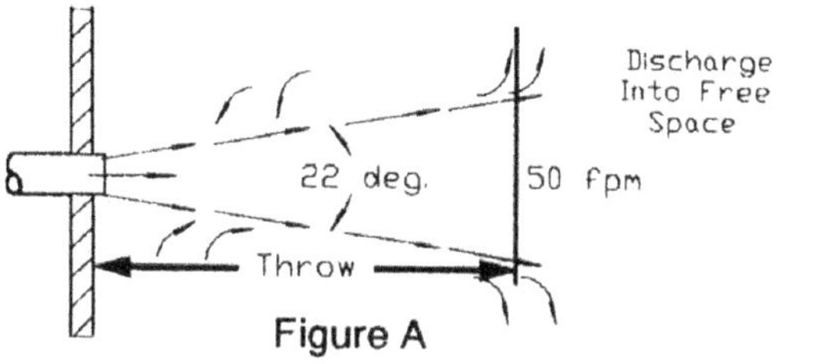

Figure A

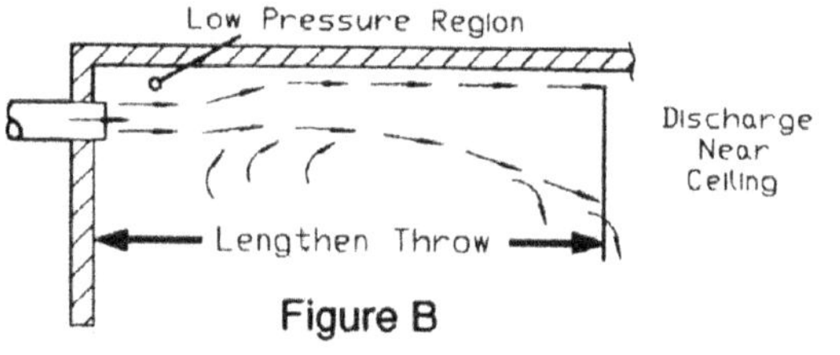

Figure B

2-9 Spread

Through the use of vertical vanes, the shape of the air stream leaving a register or grille can be adjusted to provide a narrow concentrated air pattern or spreading air pattern. Three typical examples of air patterns are provided in the illustrations below. Note that with a straight (Zero degree deflection) vane setting, the air pattern spreads like a free jet and the width or spread of the air pattern increases as the angle of the vane setting increases. Also note that the throw decreases as the spread increases.

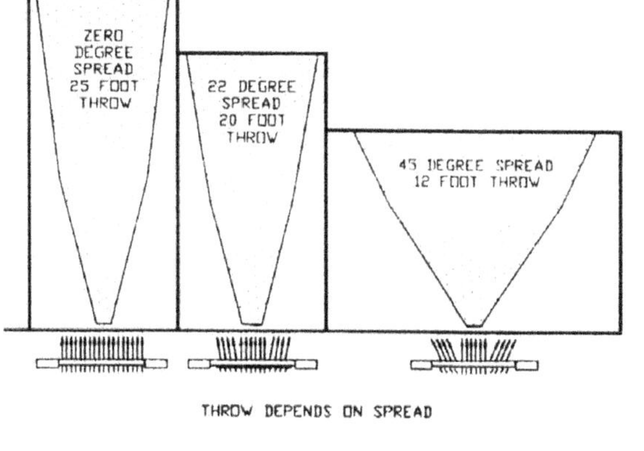

THROW DEPENDS ON SPREAD

2-10 Primary Air Pattern

The primary air pattern is characterized by an envelope of relatively high velocity air that consists of a mixture of raw supply air and entrained room air. Velocities within this envelope are equal to or greater than 150 fpm.

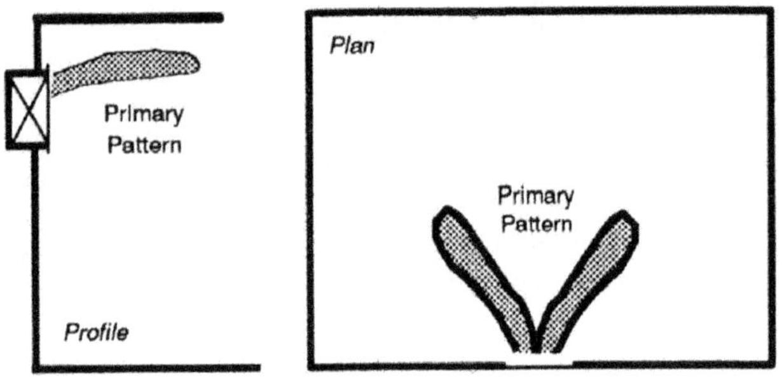

Temperatures within the primary air pattern are noticeably different than the room temperature. The primary air pattern is essentially the same for heating or cooling because the temperature difference between the supply air and the room air has no effect on the shape and behavior of the primary air pattern.

2-11 Total Air Pattern

The total air pattern is characterized by the envelope which encloses all the air that is set in motion by the primary jet and which is moving at velocities below 150 fpm. (The total air pattern does not include air that is in motion because of the effect of convective currents.)

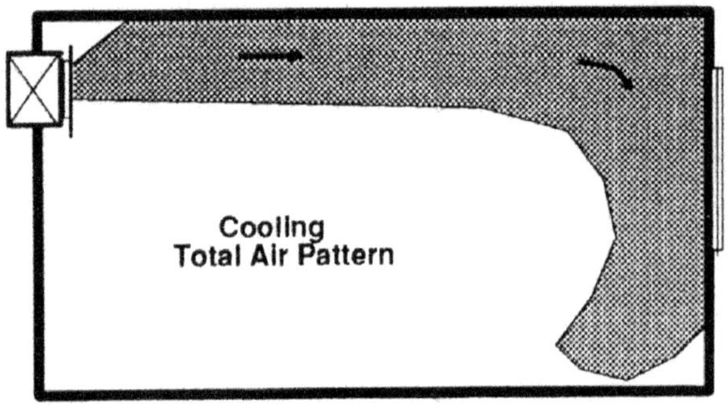

The shape and behavior of the total air pattern is affected by the supply air temperature (refer to the preceding and the following diagrams). During cooling the total air pattern tends to drop and during heating the total air pattern tends to rise or to hang near the ceiling. Also note that the total air pattern tends to cling to parallel wall or ceiling surfaces.

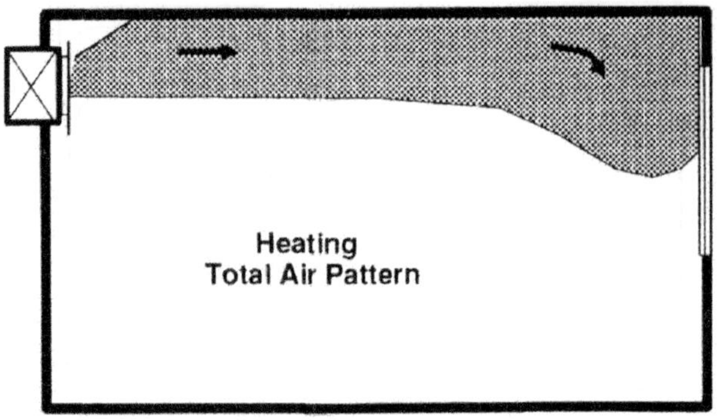

2-12 Entrained Air

Entrained air consists of the room air that drifts toward and becomes entrained in the faster moving primary and total air patterns. The figure below illustrates the relationship between the total air pattern, the zone of entrained air and the stagnate zone.

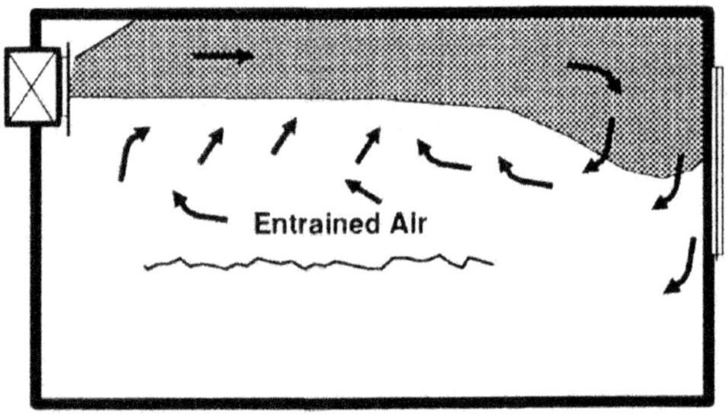

2-13 Stagnate (or Stratified) Zone

The air within the stagnate zone is not at all affected by the primary air pattern or the total air pattern. Air motion within the stagnate zone is typically less than 15 fpm and this movement (if any) is caused by natural convection currents.

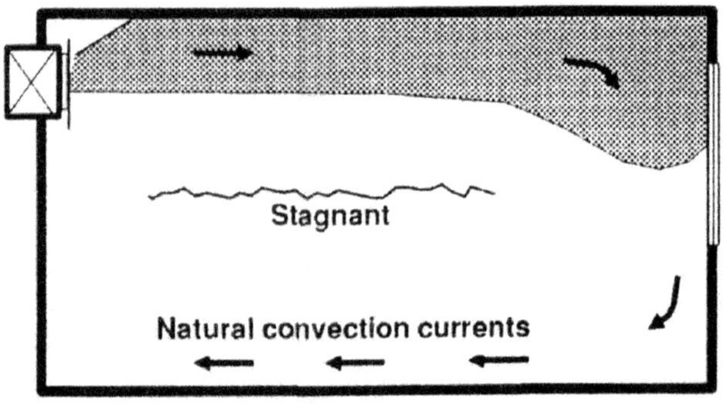

2-14 Vertical Temperature Gradient

The vertical temperature gradient is defined by the change in temperature that corresponds to an incremental, say one foot, change in vertical position. No temperature gradient occurs when supply air and room air are thoroughly mixed. Temperature gradients can only occur when poor mixing results in a zone of stagnate air. In other words, the presence of a temperature gradient always indicates the presence of a zone of stagnate air.

The vertical temperature gradient, is affected by the quality of construction and by the performance of the supply outlets.

- Larger vertical temperature differences (gradients) are associated with poorly constructed and/or poorly insulated homes (or rooms) and smaller gradients are associated with tight, well insulated homes (or rooms).
- Larger gradients are associated with supply outlets that do a poor job of mixing and diffusing the supply air with the room air and smaller gradients are associated with supply outlets which thoroughly mix the supply air with the room air.

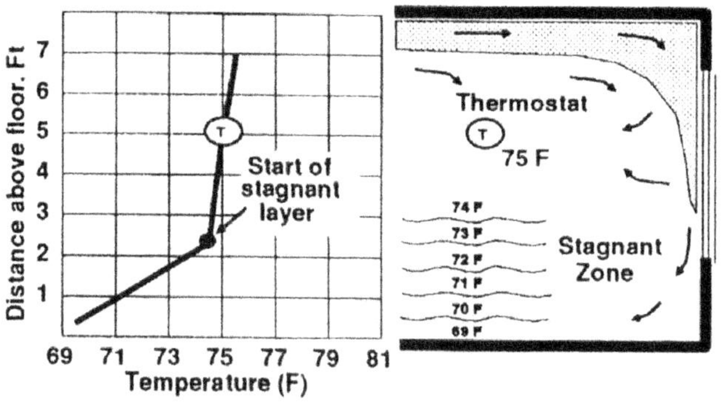

Vertical temperature gradients are often used to analyze supply outlet heating performance. The diagram below shows that the floor-to-ceiling temperature difference increases with decreasing (colder) outdoor temperatures, (or increasing heating load). Note that the floor-to-ceiling temperature difference is 5.9 F when the outdoor temperature is -5 F and only 4 F when the outdoor temperature is 25 F.

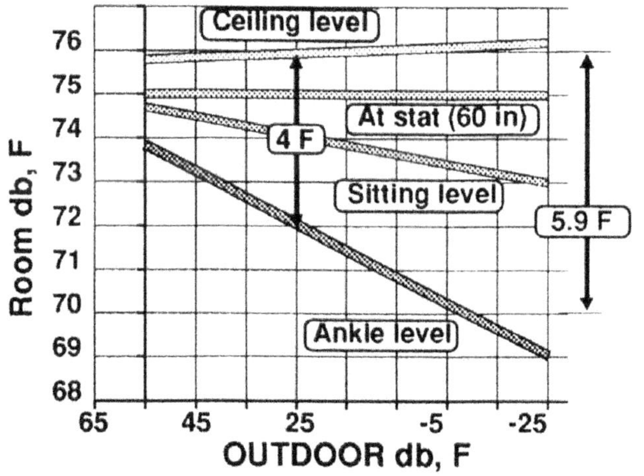

2-15 Gross Area

Gross area is the area that is associated with the inside dimensions of the frame. Face velocity is based on the gross area.

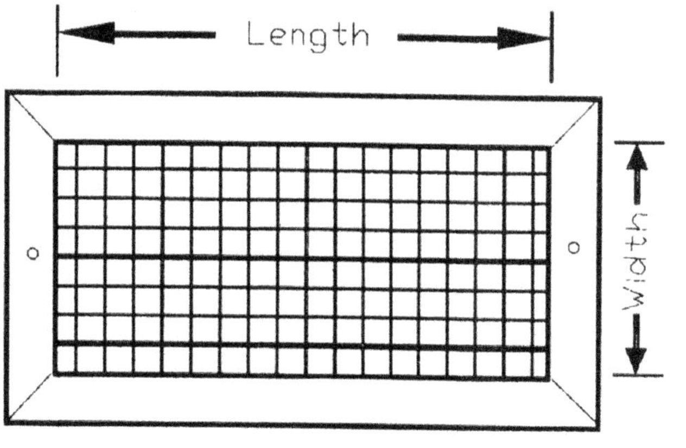

2-16 Free Area

The free area of a register or grille is the minimum opening "daylight area" available for the air to pass through the vanes. The free area determines the velocity of the air that is leaving the terminal. The free area also affects the pressure drop across the terminal. Free area typically varies from about 75 percent to about 90 percent of the area that is associated with the inside dimensions of the frame. Refer to the manufacturer's data for the exact value.

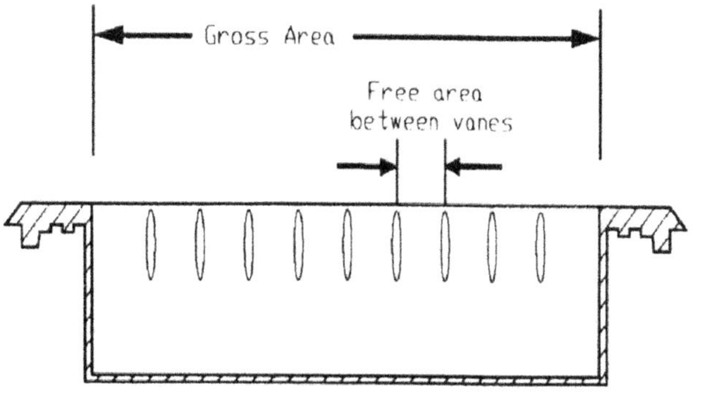

2-17 Face Velocity

Face velocity is the velocity of the air stream when it is measured within an inch or so from the face of the outlet or inlet. Face velocity can be estimated by dividing the flow rate (CFM) by the gross area (Sq. Ft.) of the inlet or outlet.

2-18 Core Velocity

Core velocity is the velocity the air stream would have if it is measured between the vanes, blades, louvers etc. Core velocity can be estimated by dividing the flow rate (CFM) by the core area of the inlet or outlet. (The core velocity can also be estimated by using the Ak factor. Refer to section 12-2.)

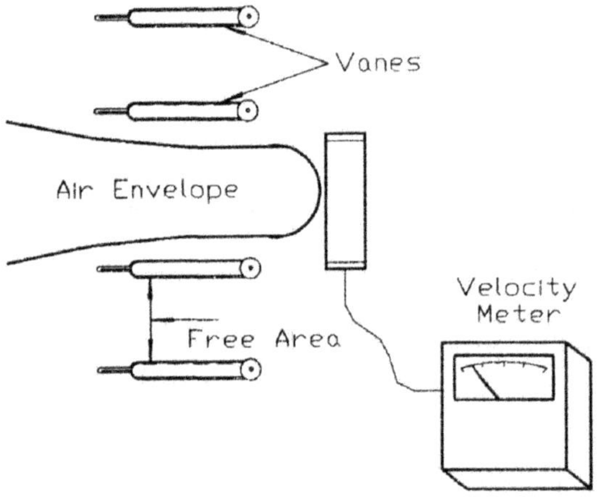

2-19 Velocity/Flow-Rate Equation

The air flow rate (CFM), the cross sectional area (of the grille or duct) and the velocity (flowing through the grille or duct) are related by the following formula.

$$Q = A \times V$$

Where:

- Q is the CFM
- A is the cross-sectional area in square feet .
- V is the air velocity in feet per minute

Note that the area and velocity can usually be measured in the field and once they are measured, the flow rate can be calculated.

Suppose a pitot tube and a manometer was used to determine the average velocity in a supply trunk with the following results:

- Velocity = 780 fpm
- Duct width = 26 inches
- Duct height = 14 inches

Then:

Area in Sq.Ft. = (26 x 14)/144 = 2.53

and ...

CFM = 780 x 2.53 = 1974

(A similar procedure can be used for grilles and registers.)

Section 3

High Sidewall Terminals

This section will discuss the air patterns that can be expected to occur when high side wall terminals are used. These air patterns result from the interaction of the primary supply air that is discharged by the high side wall terminal with the room air. The air patterns are also affected by the vertical and horizontal vane settings and the position of the terminal relative to the ceiling.

3-1 Location

In structures that are well insulated, high sidewall registers can be located on either the inside (interior) or the outside (exterior) walls. However, the interior location is preferred because the supply ducts are not easily installed at the outside wall and because heating performance is better when the supply air is thrown toward the outside wall. As will soon be apparent, a highwall location is effective for cooling and supplying ventilation air but is less effective during the heating condition.

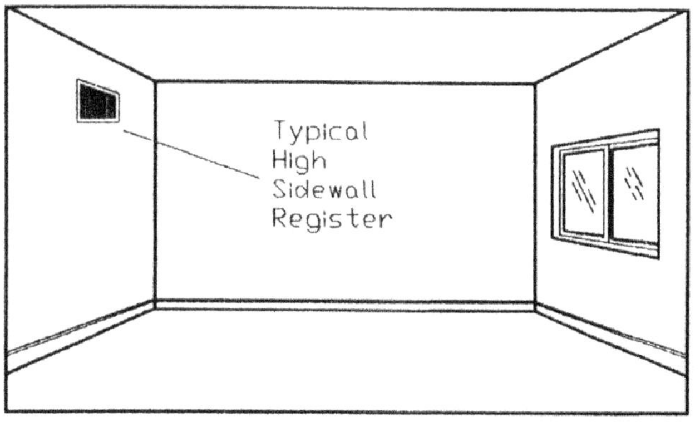

3-2 Primary Air Pattern

The primary air pattern is located close to the the high sidewall register. The primary air pattern (which includes a mixture of supply air and some entrained room air) has a velocity of 150 fpm or greater. The profile of the primary air (illustrated below) is unchanged for heating and cooling. The mixed temperature of the air in the primary air pattern varies from 3 to 4 degrees below room temperature during cooling and 20 or so degrees above room temperature during heating (assuming conventional gas or oil-fired furnaces).

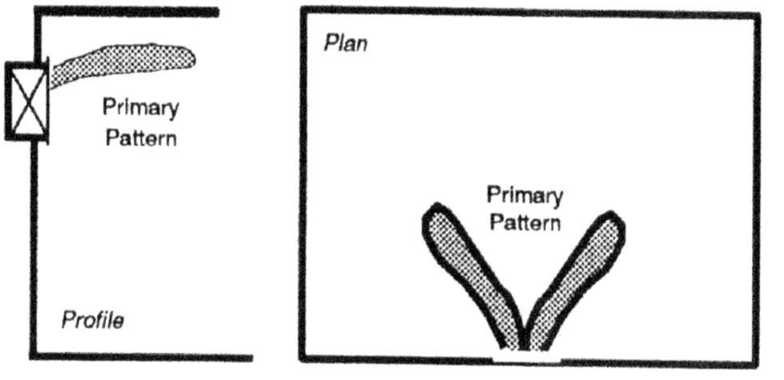

3-3 Entrance Condition Affects the Air Pattern

The diagram below shows a side wall terminal with a short collar attached to the side of a supply duct. (Note that there are no devices to help the air turn toward the terminal and then make a uniform approach to the face.) As the supply air turns to move through the diffuser the "air" hugs the right side of the collar and there is a large dead air space (separation) on the left. This type of condition destroys the symmetry of the spread pattern, has an undesirable affect on the throw and tends to increase the pressure drop through the outlet.

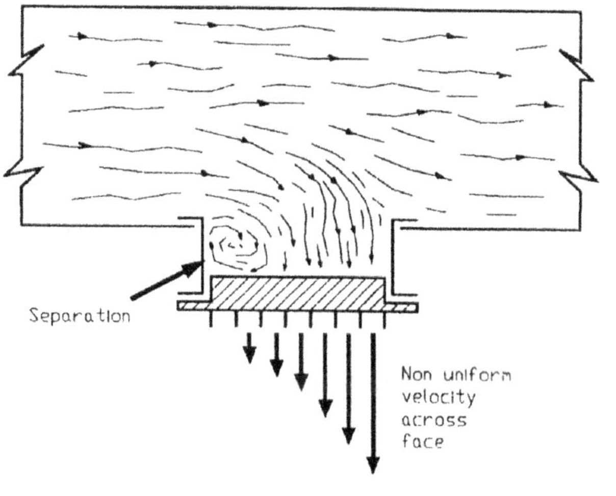

Also note that the undesirable condition illustrated in the diagram above would have been much worse if the sidewall register was installed flush with the side of the duct. ***Never install a sidewall terminal flush on the duct, always use a suitable collar.***

One method to minimize "entrance related" distortions of the air pattern is to install an extractor in the duct. The extractor helps turn the air toward the diffuser and helps distribute the flow uniformly across the face of the diffuser. This device helps assure the register will perform as tested in the laboratory.

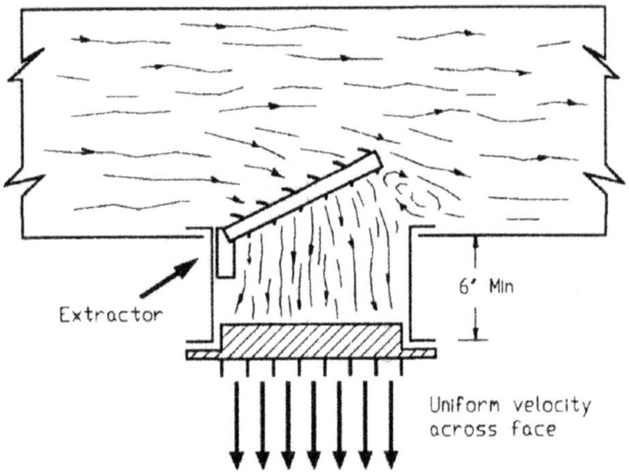

3-4 Total Air Pattern

The total air pattern is a mixture of primary air and much more room air, all at lower velocities but still under the influence of the supply outlet. Total air has a maximum velocity of 150 fpm and no sharply defined lower velocity. Unlike primary air, the temperature difference between total air pattern and room air does have an effect, so the total air pattern during heating is different than during cooling.

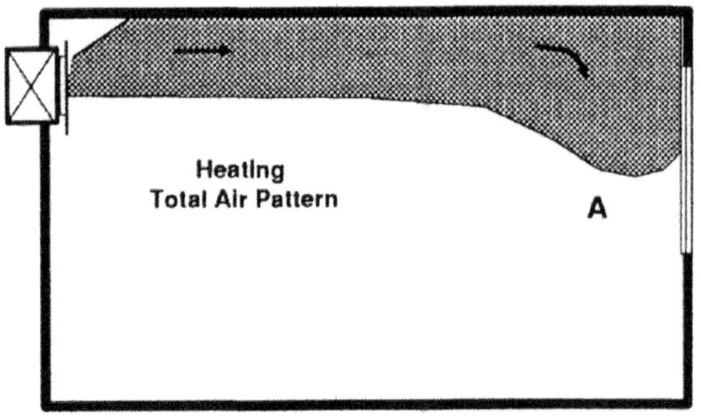

The total air pattern tends to rise during heating and tends to drop during cooling. Note the difference in total air pattern profiles in the preceding and following illustrations. During heating, the total air pattern reaches about to the middle of the window (Point A), while for cooling, total air pattern extends almost to the floor (Point B). This characteristic obviously has an effect on comfort experienced within the occupied space. Also note that the total air pattern usually moves along in contact with room surfaces (refer to Section **2-8**).

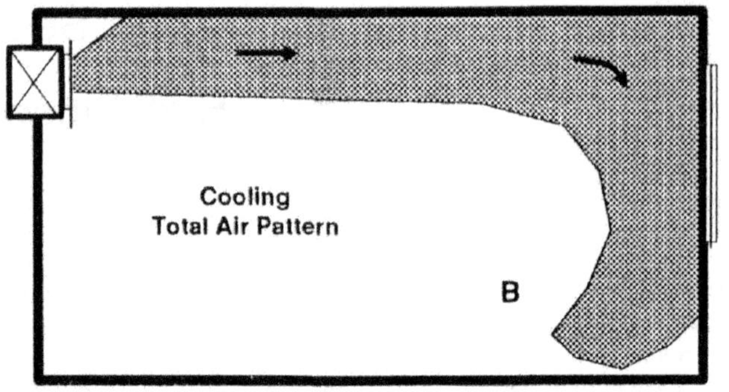

3-5 Stratification

When high sidewall terminals are used, stratification occurs during the heating season. In this case the layer of cooler air which accumulates near the floor is well beyond the influence of the total air pattern. In the figure below the arrows along the floor depict the direction of the corresponding natural convection currents.

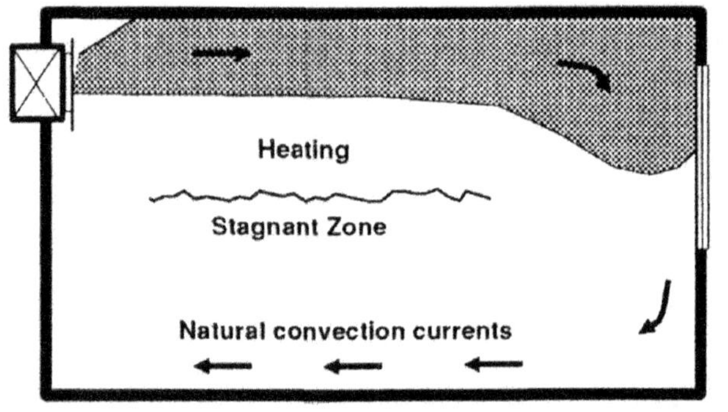

3-6 Secondary Air Pattern

The secondary air pattern is not part of the total air pattern nor is it part of the stagnant or stratified zone. This air pattern (as shown by the arrows) tends to drift back toward the primary and total air patterns and become entrained in the aspirating action of the faster moving air. This air pattern is well within the occupied zone but fortunately the velocities within this pattern are relatively low.

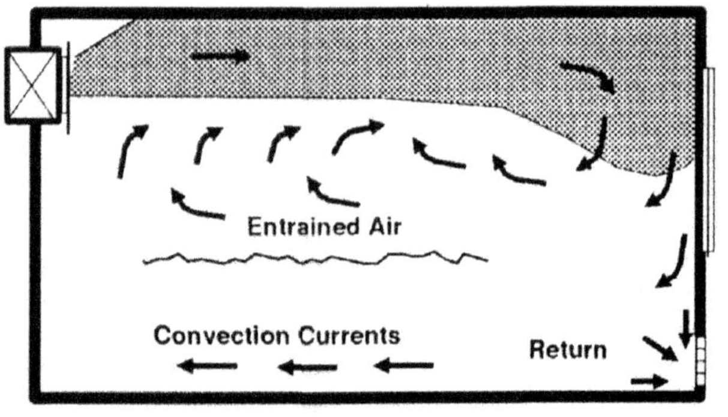

3-7 Use Vanes to Control Air Patterns

Through the use of vertical vanes, the shape of the primary and total air patterns can be adjusted to provide a narrow, concentrated pattern or a spreading air pattern. Here in plan view are three typical examples of spread patterns for three different vane settings. Note that the long narrow pattern is ideal for throwing the air across the length of a long narrow room and the wide pattern is ideal for throwing the air across the width of a short wide room.

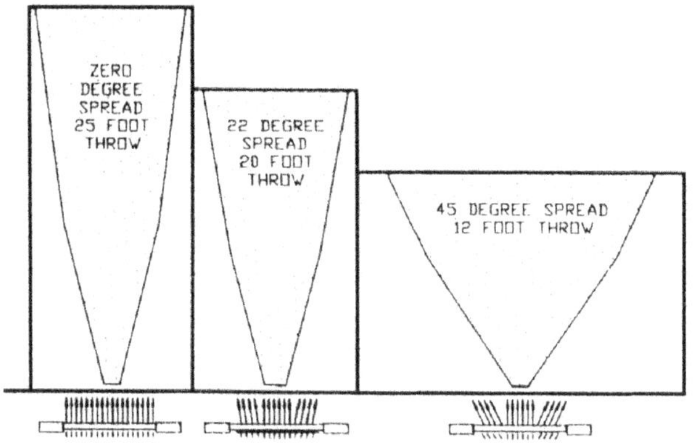

THROW DEPENDS ON SPREAD

3-8 Characteristics of High Sidewall Registers

The following comments are intended to illustrate some of the general performance characteristics of high sidewall registers. In actual practice the affects described below will vary depending on make, model, size and air flow rate.

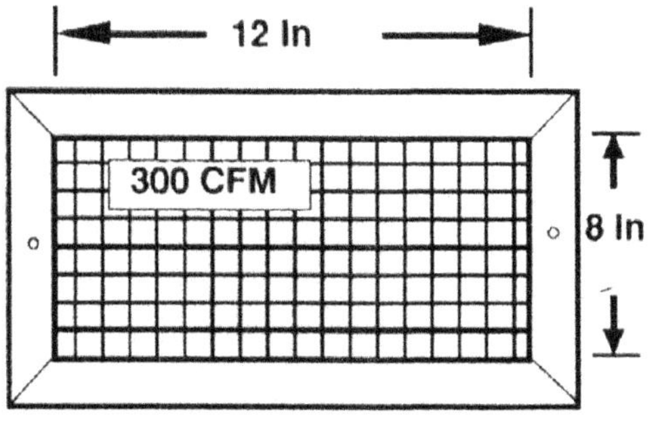

Consider a 12 inch x 8 inch high sidewall register (not near the ceiling) with double deflection vanes carrying 300 CFM of cooling air with a 20-degree supply air to room air temperature differential and a 500 fpm face velocity. (The performance of similar registers with the same core area, but different widths and heights, would be essentially the same.)

Effect of horizontal vanes setting:

In Sketch A, the horizontal vanes are adjusted to provide a 5-degree vertical (upward) deflection and the vertical vanes are set for a zero degree horizontal deflection. The throw for a 50 fpm terminal velocity is approximately 28 feet and the drop is 7 feet.

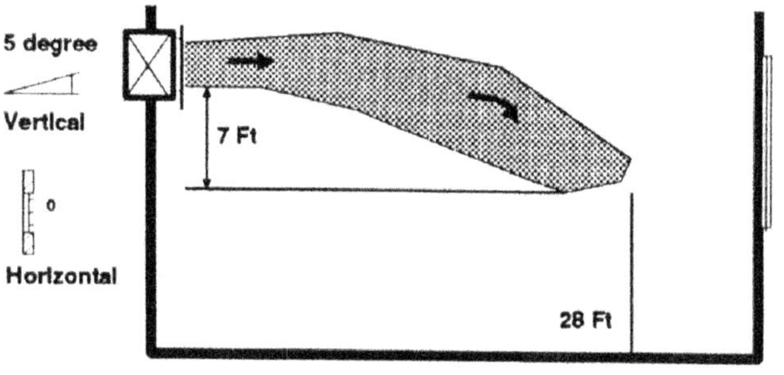

Sketch A

If the horizontal vanes are adjusted to provide a 15-degree upward deflection while the horizontal vanes remain set a zero degrees the drop is reduced to 3 feet and the throw is approximately the same. (Sketch B).

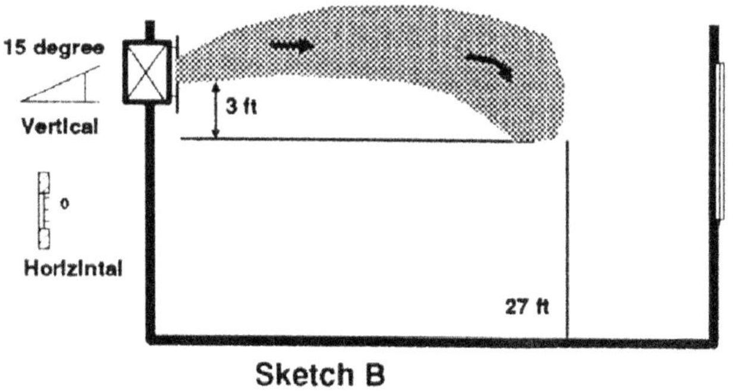

Sketch B

Effect of vertical vane setting:

In Sketch C below, the horizontal vanes are returned to a 5 degree vertical (upward) deflection position and the vertical vanes are set for a 45 degree left and 45 degree right spread. Compared with Sketch A, the increase in spread reduces the throw by 12 feet and the drop by 4 feet.

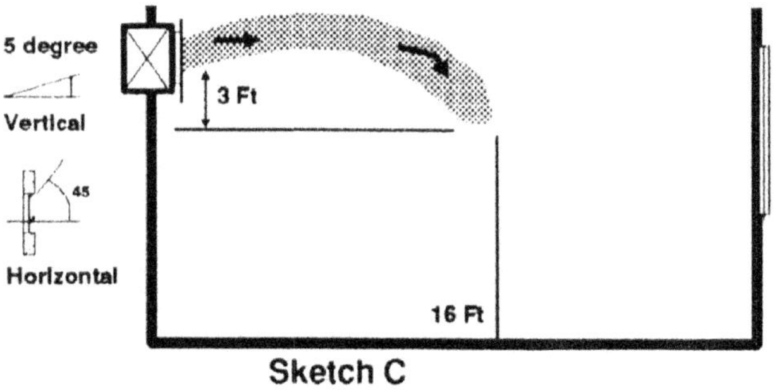

Sketch C

Effect of supply temperature:

If the supply air temperature is increased so that the room temperature to supply air temperature differential is 10 degrees instead of 20 degrees, the throw will increase and the drop will decrease. If the supply air temperature is decreased (made colder) so that the temperature differential (TD) is 30 degrees, the drop will increase and the throw will decrease (Sketch D).

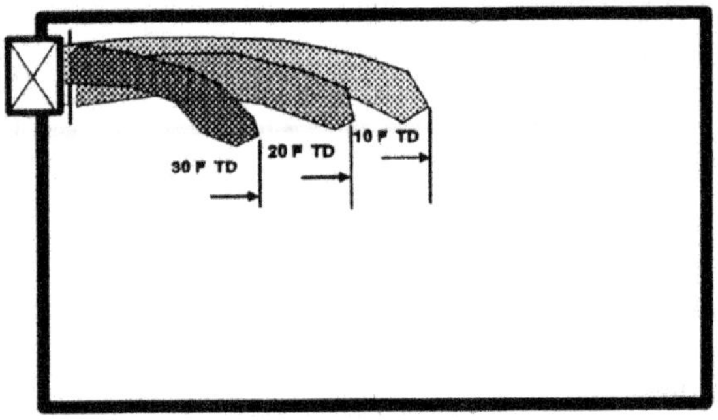

Sketch D

Effect of proximity to the ceiling:

If the register in Sketch A (5 degree vertical, Zero degree horizontal settings) is placed within 1 foot of a ceiling, the throw will be extended 3 feet (from 28 to 31) and the drop will be reduced from 7 feet to 6 feet. (Sketch E).

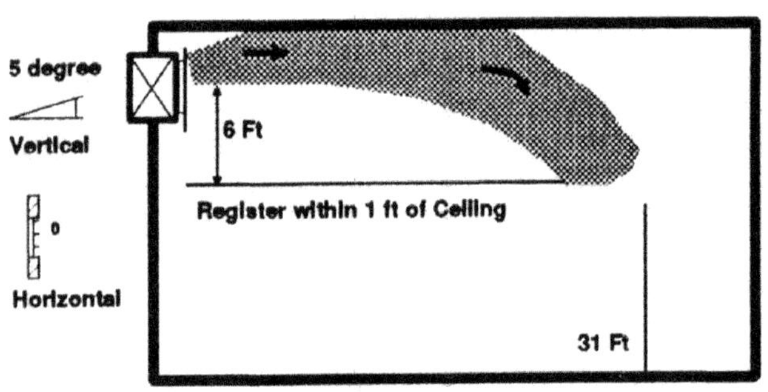

Sketch E

When a register is not within one foot of the ceiling, a vertical deflection adjustment can help obtain some ceiling effect. In Figure F below, the register is from 2 to 4 feet below the ceiling, but a 15 degree upward deflection helps initiate the surface effect which extends the throw and reduces the drop to only 2.5 feet.

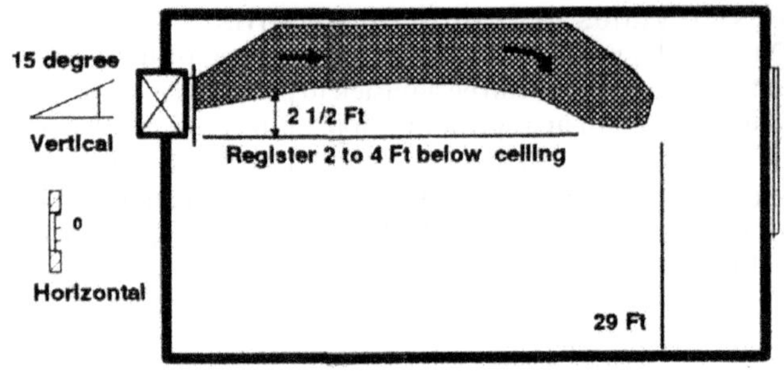

Sketch F

Also note that the ceiling (surface) effect tends to offset the tendency of a spreading pattern to reduce the throw. Compare Sketch G (located near the ceiling with a 45 degree horizontal spread) with Sketch C (not located near the ceiling with a 45 degree horizontal spread). Note that the proximity of the ceiling extends the throw from 16 feet to 23 feet.

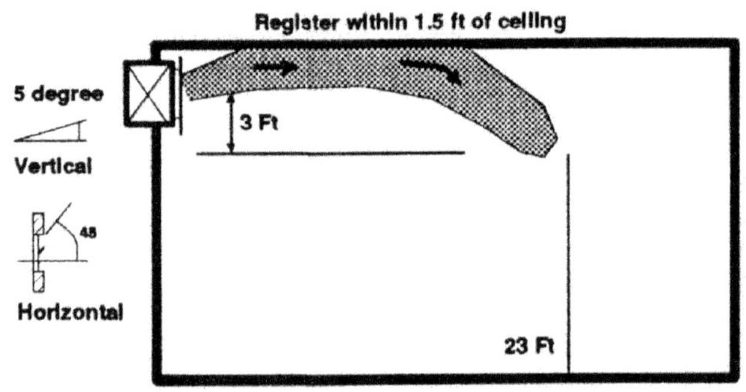

Sketch G

3-9 Location of Returns

In general the location of the return has no effect on the air pattern within the occupied zone and negligible effect on the shape and size of a stagnate zone. The figure below illustrates how the total air pattern (during heating) is influenced by the location of the return air inlet. In this instance, the return is shown placed beneath a window on the outside wall. Note that the arrows indicating the path of natural convection currents have not been substantially disturbed. This means the return located under the window would "capture" or draw in very little cold air moving down the outside wall. (Refer to section 7-2 for a discussion of why the return has so little influence.)

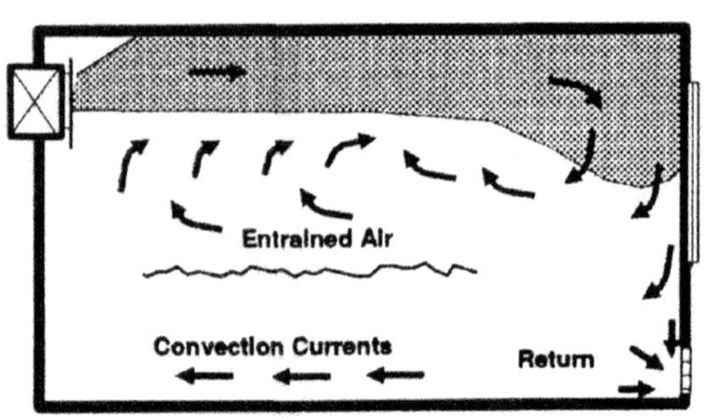

Section 4

Ceiling Diffusers

This section will discuss the air patterns that can be expected to occur when ceiling terminals are used. These air patterns result from the interaction of the primary supply air that is discharged from the ceiling terminal with the room air. The air patterns are affected by the temperature difference between the room air and the supply air and proximity of the terminal relative to the ceiling.

Ceiling diffusers can be classified as either horizontal discharge or vertical discharge.

- Horizontal discharge diffusers include all round, square, rectangular and slot diffusers which turn the primary air jet so it flows parallel to the ceiling. These diffusers can be subdivided into types that create a non-directional, radial air pattern and "directional" types which are designed to produce a one-way, two-way, three-way or four-way air pattern.
- Vertical diffusers include all round, square, linear and slot diffusers which project the air straight down toward the floor.

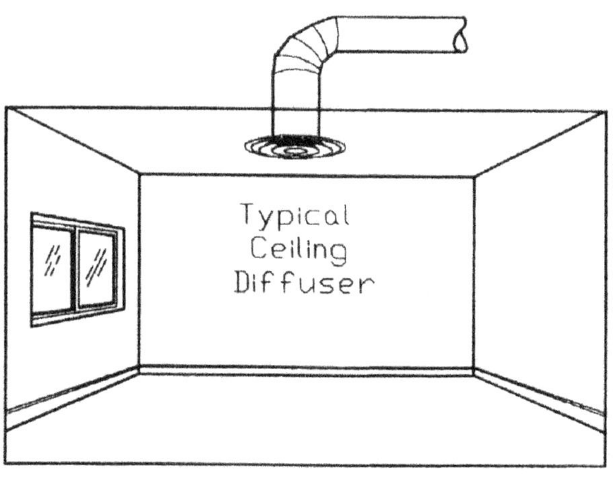

4-1 Primary Air Pattern, Radial Discharge

The primary air pattern (velocity of 150 fpm or greater) for a non-directional round, square or rectangular ceiling diffuser forms a horizontal circular pattern which surrounds the terminal. The throw, or radius, of the ceiling diffusers' primary air pattern is relatively short because a circular pattern spreads the primary air over a relatively large area. Because of the large diffusion area, the ceiling terminal can handle large air flow rates (CFM) without producing drafts.

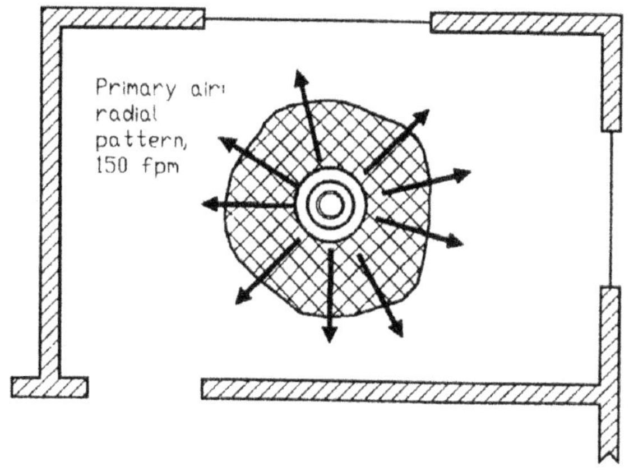

The sketch below illustrates the "flow lines" that are created when a ceiling diffuser discharges into a room. Note the right and left discharge as the primary air jet as it leaves the diffuser. Also note that (initially) the primary air jet is not horizontal with the ceiling. At the bottom of the sketch, room air is moving up to be entrained by the supply jet, demonstrating quite effectively the induction characteristic of a ceiling diffuser. Because of this characteristic, the air quantity under the influence of the diffuser can be 10 times greater than the original supply CFM.

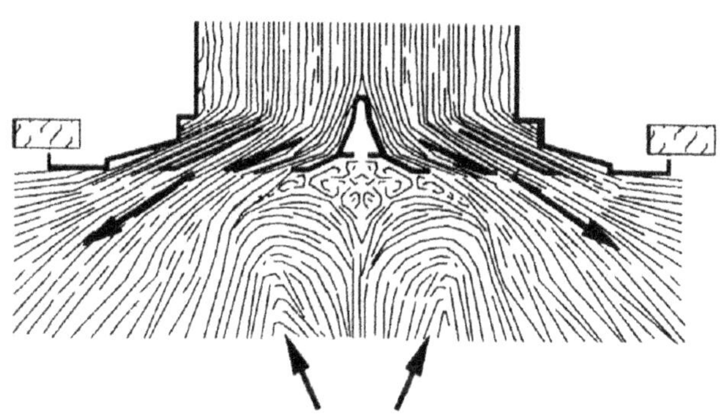

4-2 Primary Air Pattern, Directional Discharge

Directional ceiling diffusers are designed to provide one-, two-, three- and four-way air patterns. In general, at a given air flow rate, the throw increases as the air pattern becomes more directional. For example, the throw of a two-way pattern will be greater than that of a three-way pattern. The ability to adjust the air pattern (especial when the adjustment can be made in the field) adds greatly to the versatility of these types of diffusers. Ceiling diffusers with one-way blow toward the window and outside wall generally provide improved heating performance over a (non-directional) diffuser with a 360 degree pattern.

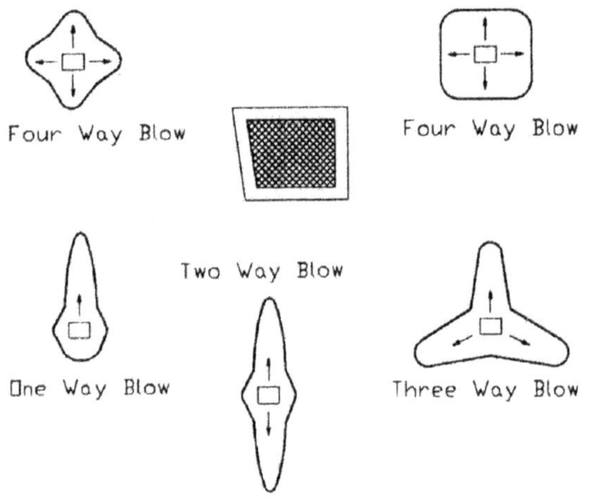

4-3 Primary Air Pattern, Vertical Discharge

Ceiling mounted outlets featuring a vertical (downward) discharge usually create a primary air pattern that is similar to a non-spreading jet of air. Long linear and long slot diffusers create a vertical curtain of air which also has little tendency to spread. Note that in both cases the shape of the primary air pattern is similar during heating or cooling, but during cooling, the primary air pattern penetrates further down into the room.

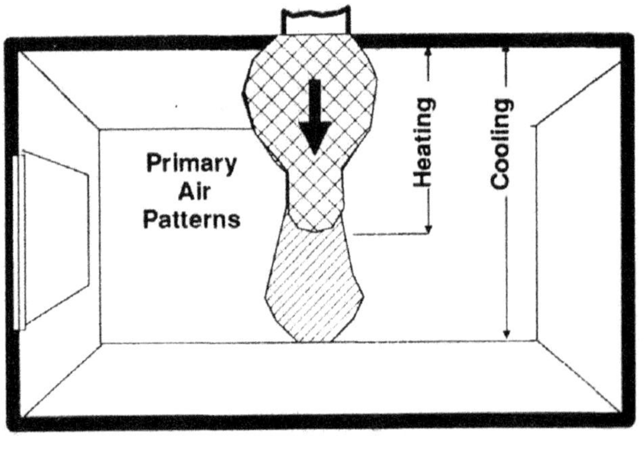

4-4 Entrance Conditions Affect the Air Pattern

The sketch below shows a diffuser with a short collar attached to the bottom of a supply duct. (Note that there are no devices to help the air turn toward the diffuser and then make a uniform approach to the diffuser.) As the supply air turns to move through the diffuser the air hugs the right side of the collar and there is a large dead air space (separation) on the left. This type of condition destroys the symmetry of the air pattern, adversely affects the radius of diffusion and tends to increase the pressure drop through the outlet.

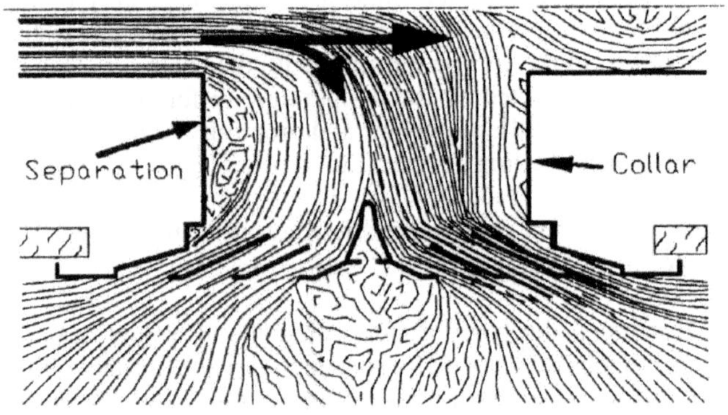

Also note that the undesirable condition illustrated in the sketch above would have been much worse if the ceiling diffuser was installed flush with the duct. ***Never install a ceiling diffuser flush with the bottom of the duct, always use a suitable collar.***

One method to minimize "entrance related" distortions of the air pattern is to install an extractor in the duct and a straightening or equalizing grid at the throat of the collar. The extractor helps turn the air toward the diffuser and the grid distributes the flow uniformly across the neck area. These devices help assure diffuser will perform as tested in the laboratory.

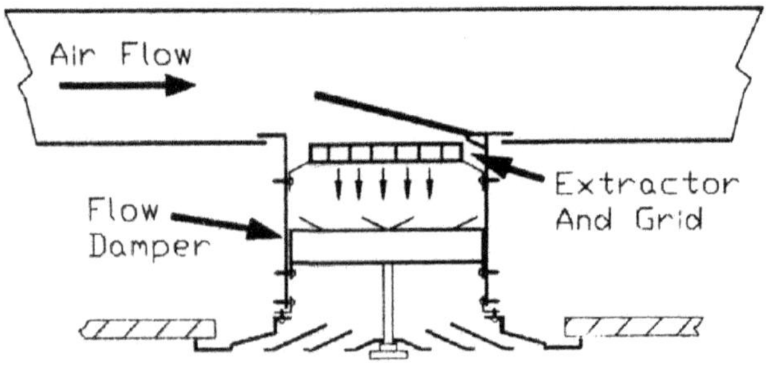

4-5 Total Air Pattern, Horizontal Discharge

During cooling, the total air pattern that is associated with horizontal discharge diffusers consists of an envelope of primary air and induced room air. This envelope will move further down on the inside wall (A) than it will on the outside wall, which has a window (B). This difference occurs because relatively strong natural convection currents moving up along the window counteract the flow of the total air pattern and this tends to cause the window side of the pattern to drop before it reaches the outside wall.

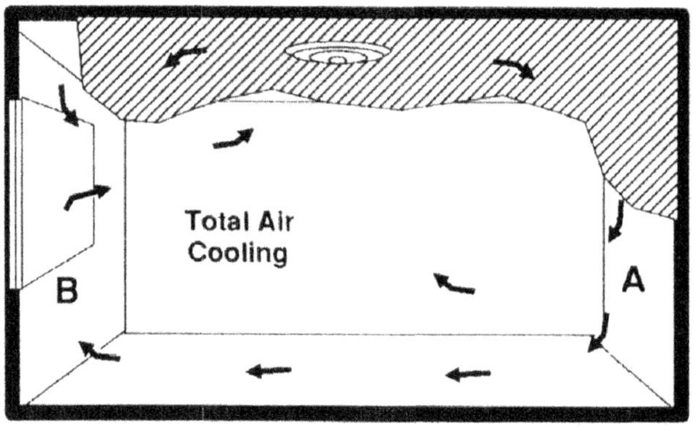

Also note that during cooling the warmest room air, which is near the ceiling, is the first to make contact and mix with the primary air. This mixing reduces the temperature of the primary air pattern and this tends to increase the diffusion capability of the terminal.

During heating, the total air pattern consists of an envelope of buoyant air which tends to cling to the ceiling. This envelope will move further down on the outside wall, which has a window (C), than it will on the inside wall. This difference occurs because of the relatively strong natural convection currents moving down the window tend to "pull" the edge of the total air pattern down. Therefore, during heating, slightly more total air reaches the cold window and outside wall (C) than reaches the inside wall (D).

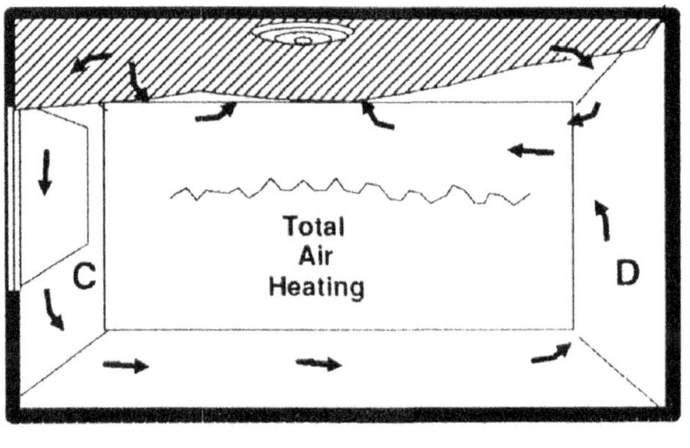

Note that horizontal discharge ceiling diffusers are exceptionally poor at heating below grade rec-rooms. The diagram below shows that when they are used for this application the stagnant zone extends to more than five feet above the floor and that the floor surface temperatures are relatively cold everywhere.

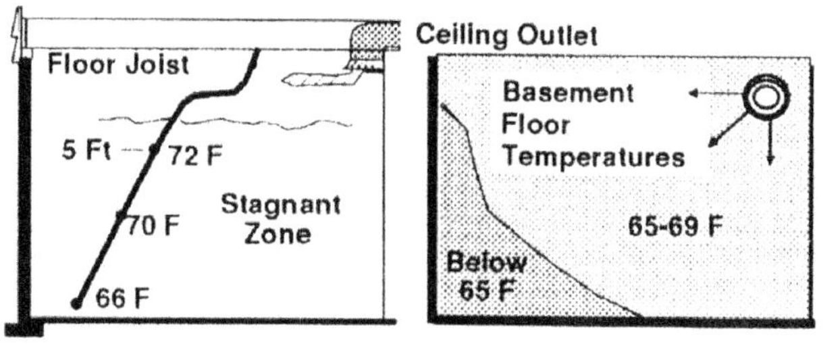

4-6 Total Air Pattern, Vertical Discharge

Square and round ceiling diffusers featuring vertical discharge create a cylindrical pattern which tends to mushroom out near the floor. Long linear and long slot diffusers create a vertical curtain of air which also tends to spread out near the floor.

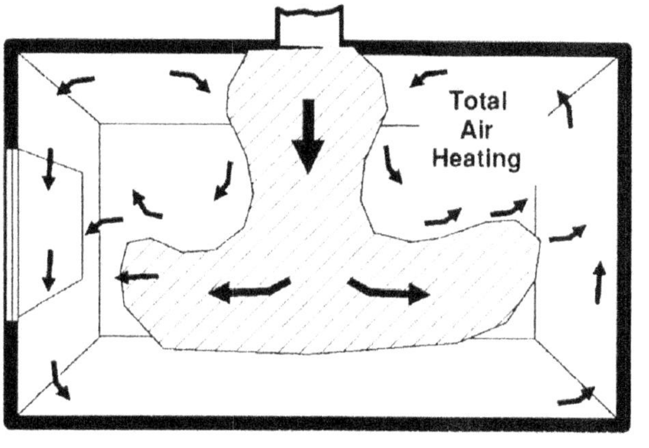

During heating, round or square diffusers located in the interior of the room can provide a desirable total air pattern but drafty conditions will occur if the jet of primary air penetrates into the occupied zone. Therefore, the benefit of using higher jet velocities to get the air down to the floor is offset by the tendency of the higher velocities to create drafts. For this reason these types of diffusers are only recommended when comfort considerations are not of primary concern, such as a warehouse heating application.

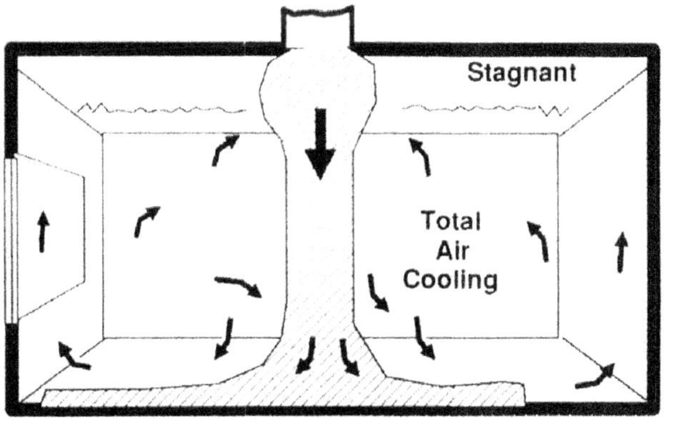

Cooling with downward projecting ceiling diffusers results in a concentrated total air pattern that produces a pool of cold air near the floor. It is also probable that unacceptably high velocities will occur in the occupied zone. Also note that when a downward projecting linear or slot diffuser is located close to a wall, the air pattern will tend to cling to the wall and the throw will be increased. Long linear and slot diffusers installed next to an outside wall create a blanket of supply air which tends to offset the wall and window loads. Diffusers located close to the wall have the advantage of creating a downward projecting air pattern that is outside of the occupied zone.

4-7 Stratification, Horizontal Discharge

During heating, the total air pattern that is associated with horizontal discharge diffusers tends to cling to the ceiling and a large stagnant region develops below. This can cause a very noticeable temperature differential between the floor and the ceiling. In cold climates the combination of slab construction and ceiling terminals may result in unsatisfactory comfort at the foot/ankle level.

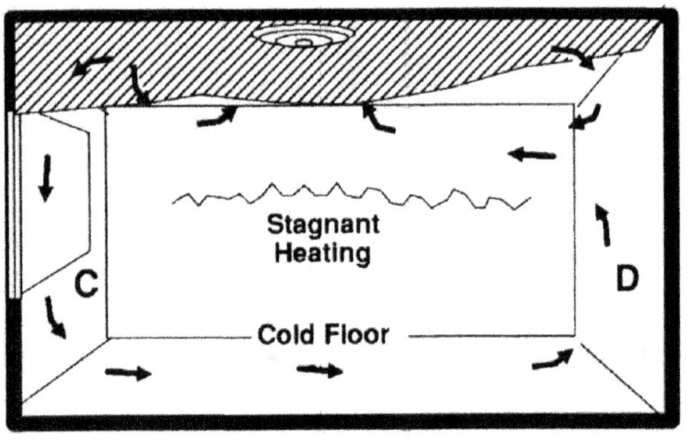

4-8 Stratification (Vertical Discharge Diffusers)

During heating, the stagnant region will be located near the floor and will be negligible if the diffuser is sized so that the total pattern air reaches down to the floor. During cooling, non-spreading downward projecting ceiling diffusers produce a stagnant region near the ceiling.

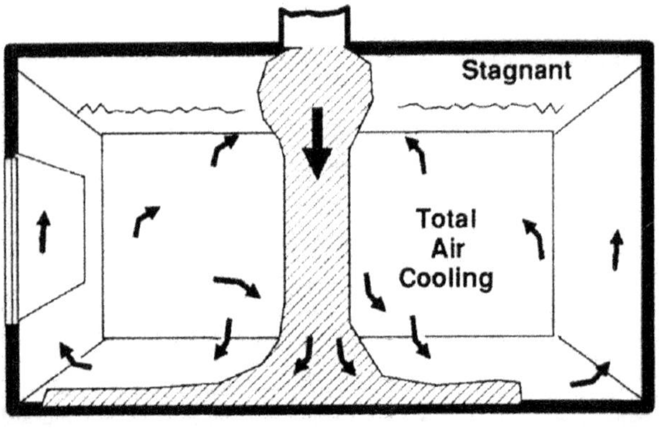

4-9 Characteristics - Horizontal Discharge Outlets

The following comments are intended to illustrate some of the general performance characteristics of non-directional and directional horizontal discharge ceiling outlets. In actual practice the effects described below will vary depending on make, model, size and air flow rate.

Effect of cooling temperature difference:

During cooling, the temperature difference (TD) between supply air and room air can have a significant effect on throw and drop. Refer to the ceiling diffuser air patterns illustrated in the figure below. With no TD, that is, the supply air and room air are at the same temperature, the throw radius for the diffuser is 15 feet. There is essentially no drop. When the TD is 30 degrees, meaning the supply air temperature was 30 degrees cooler than room temperature, the throw is reduced to 10 feet and the drop increases to 5 feet. Thus, the sup-ply air temperature has a very significant effect on outlet performance. When using manufacturer's catalog data, it is important to verify that the temperature difference that is associated with the published data corresponds to the temperature difference that is being used for the design.

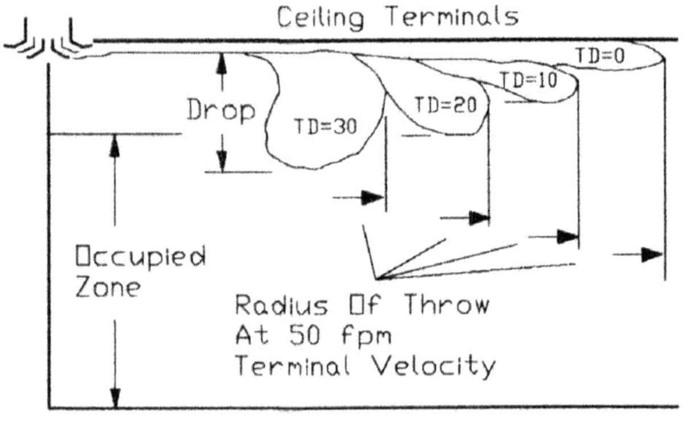

Effect of the ceiling:

The performance of ceiling diffusers is always affected by the proximity of the ceiling. With some ceiling diffusers, (especially flush style diffusers), mounting the diffuser more than 3 inches below the ceiling will cause the jet of primary air to discharge at an angle which slopes down and away from the ceiling. Lowering the mounting even further increases the angle between the ceiling and the primary air jet.

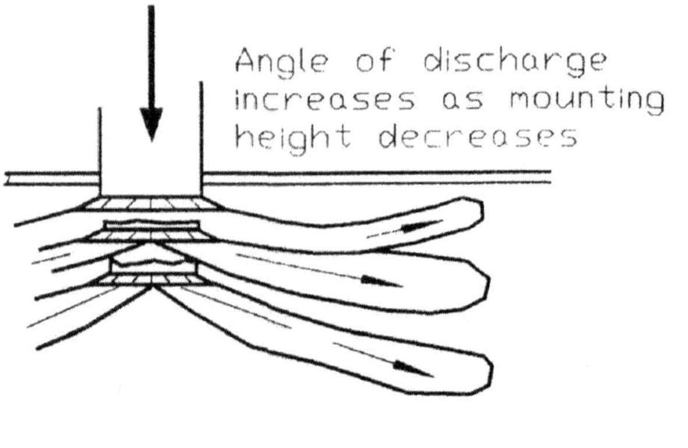

This change in discharge pattern occurs because air is induced into the space between the ceiling and primary jet, preventing the primary air pattern from "attaching" itself to the ceiling. Throw is always reduced in these instances, perhaps by as much as 25 or 30 percent, when compared to the flush mounted installation. Some manufacturers provide "free space" performance data or correction factors for "no ceiling" installations in their catalogs.

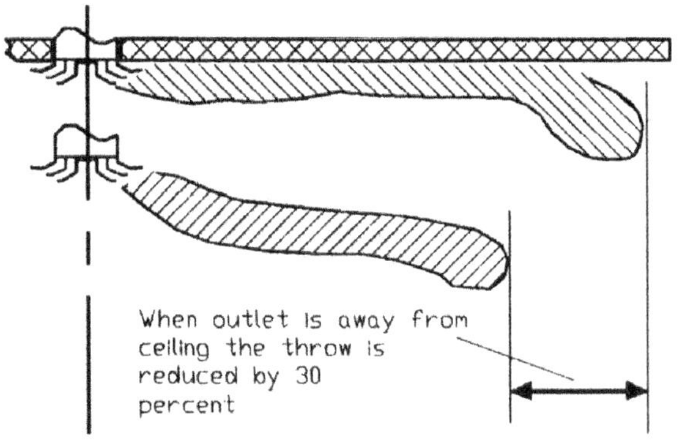

Also note that some diffusers are specially designed to provide horizontal or upward projection when there is no ceiling. (For example, when the ceiling diffusers are attached to drops from ducts located in "high bay" areas.)

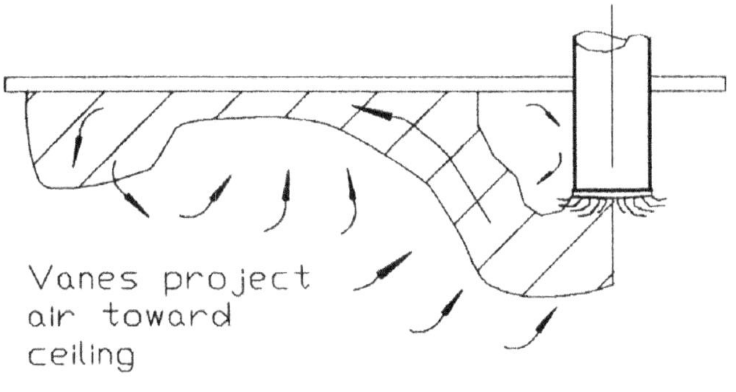

Effect of beamed ceilings:

When horizontal discharge ceiling diffusers are installed in beamed ceilings the outlets should be located below the bottom of the beam level. As discussed in the paragraph above, the throw may be reduced and the drop may be increased because the surface effect will have little or no influence when the outlet is located away from the ceiling.

Smudging tendencies:

The induction of room air by the ceiling diffuser is the reason why surfaces around a diffuser may get dirty over time, (assuming the supply air is properly filtered). As the induced room air, which may carry dust and dirt particles in suspension, sweeps past the wall or ceiling surface it deposits some of the suspended particles on the surfaces.

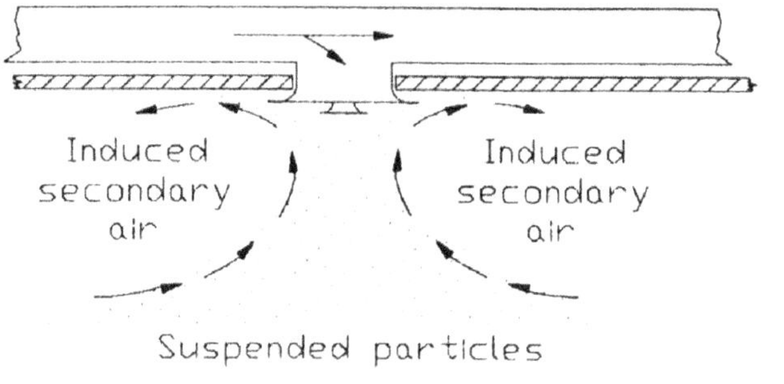

Effect of turning the air with curved vanes:

The next sketch shows how the supply air separates or lifts off the curved surfaces of ceiling diffusers' vanes. In a sense this separation increases the tendency of the vanes to obstruct the flow of supply air. The fact that the supply air does not make full use of the open area means that measurements of the collar size or of the free area are not representative of the effective flow area. The effective area (or an area factor) is usually determined by laboratory tests and is normally published in the manufacturer's performance data.

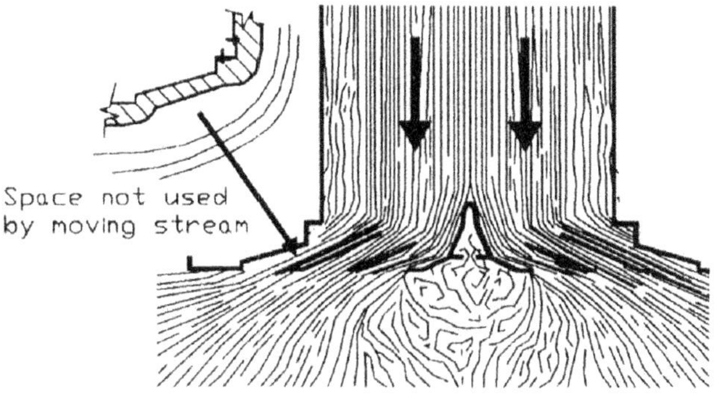

4-10 Characteristics of Vertical Discharge Outlets

The following comments are intended to illustrate some of the general performance characteristics of vertical discharge ceiling outlets. In actual practice the effects described below will vary depending on make, model, size and air flow rate.

Effect of temperature difference:

During cooling, the performance of vertical discharge ceiling terminals is not very sensitive to the changes in the supply air temperature. This is also the case during heating. However, during cooling the primary air pattern can be expected to project further down toward the floor than it would during heating.

Effect of downward vertical projection:

During heating or cooling, vertical discharge ceiling terminals which are located toward the interior of the room have the potential to cause unacceptably high air velocities within the occupied zone. However this is not necessarily a problem when linear or slot diffusers are positioned to blow down along an outside wall.

4-11 Location Of Returns

In general the location of the return has no effect on the air pattern within the occupied zone and negligible effect on the shape and size of a stagnate zone. (Refer to section 7-2 for a discussion of why the return has so little influence.)

Section 5

Floor and Low Sidewall Outlets

This section will discuss the air patterns that are associated with floor mounted diffusers, low side wall diffusers, floor registers and low side wall registers. These air patterns result from the interaction of the primary supply air that is discharged from the diffuser or register with the room air. Diffusers are characterized by a wide spreading primary air pattern while registers are characterized by relatively narrow primary air pattern. Floor and low sidewall outlet air patterns can be affected by windows, sliding glass doors and draperies.

The figure below illustrates the various types of floor and low side wall outlets. In general floor outlets are preferable to low side wall outlets and vertical discharge floor outlets are preferrable to horizontal discharge floor outlets. Inside wall loactions are generally acceptable in warm climates. In cold climates, the outside wall location is generaly preferred. However, in cold climates the inside wall location is acceptable if there are no slab or exposed floors but, the envelope must be well insulated and it must have tight thermal pane windows.

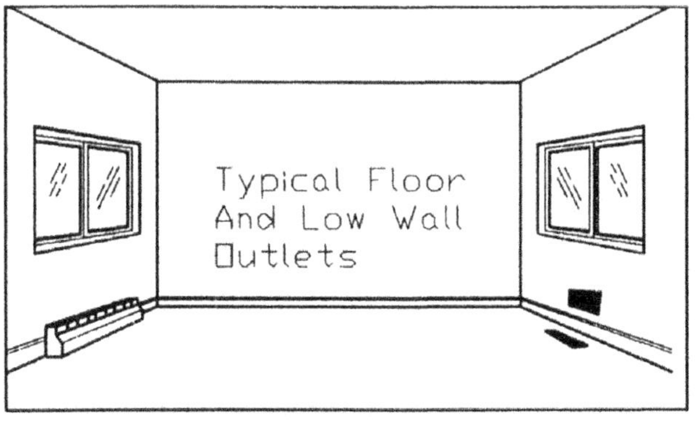

Floor and baseboard diffusers and registers are usually installed along outside walls and preferably beneath windows. Low sidewall diffusers and registers are often located on the inside wall of a room but can be located along outside walls as well. All of these types of outlets can be further classified as either vertical discharge or horizontal discharge.

- Vertical discharge outlets include all floor, baseboard, linear and low sidewall diffusers, and registers that are designed to project the primary air jet vertically along the wall.
- Horizontal discharge outlets include all floor, baseboard, and lowside wall diffusers and registers that are designed to project the primary air jet parallel to the floor or at an angle relative to the floor.

5-1 Primary Air Pattern, Vertical Diffuser

While the detailed shape of the primary air jet for each vertical discharge diffuser type is unique, each can be broadly characterized as providing wide, spreading vertical jets. The following diagrams illustrate the general shape of the primary air pattern produced by the various types of vertical discharge diffusers.

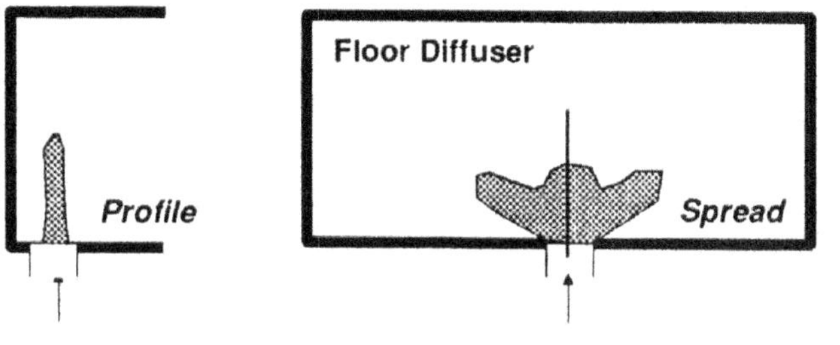

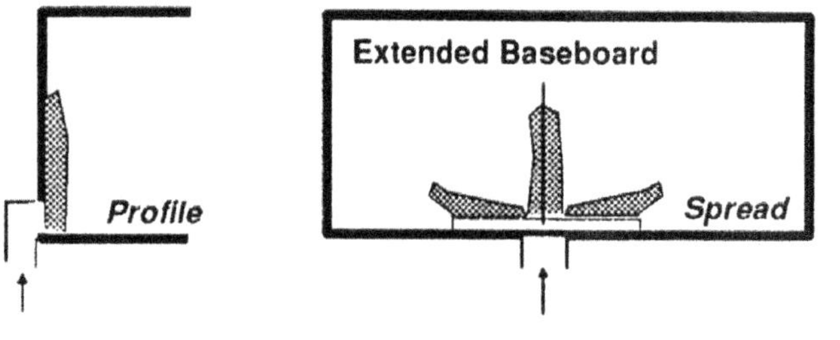

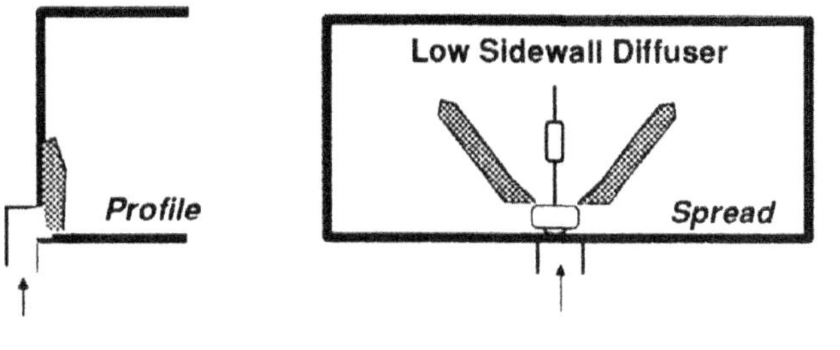

5-2 Primary Air Pattern, Vertical Registers

Vertical discharge floor registers, short baseboard registers and linear grilles perform in a manner that is similar to vertical discharge diffusers. However registers differ from diffusers in that registers feature a single, non-spreading primary air pattern as shown in the diagram below. Compare the figures below with the figures above, note the difference between the register air pattern and the patterns that are produced by the floor diffuser and the extended baseboard diffuser.

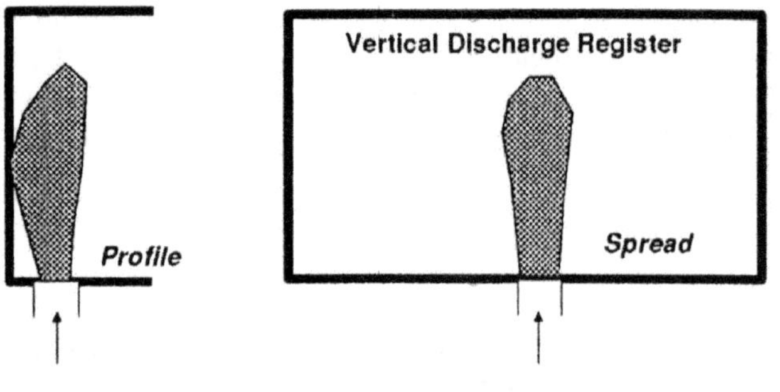

5-3 Primary Air Pattern, for Horizontal Discharge

Again, the shape of the primary air jet depends on the details of the outlet design but in general air patterns can be characterized as providing a spreading horizontal jet. The following diagram illustrates the general shape of the primary air pattern produced by horizontal discharge outlets.

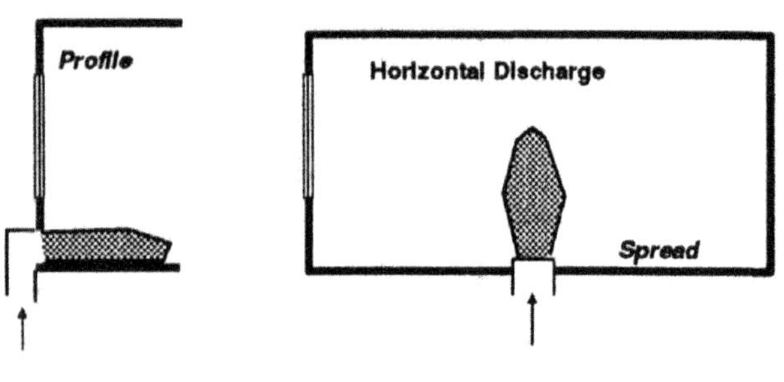

5-4 Entrance Conditions Affect the Air Pattern

Ideally the air should approach the diffuser in a direction perpendicular to the face of the diffuser and the velocity of the approaching air should be distributed uniformly across the neck of the diffuser. Various fittings and other devices are designed to obtain the desired entrance condition. These devices help assure diffuser will perform as tested in the laboratory.

Special boots and fittings are available for use with floor or linear outlets. Refer to **ACCA Manual D**, Group 7, boot fittings for some examples of boot designs. Also refer to the manufacturer's catalog for linear diffuser boot or plenum designs.

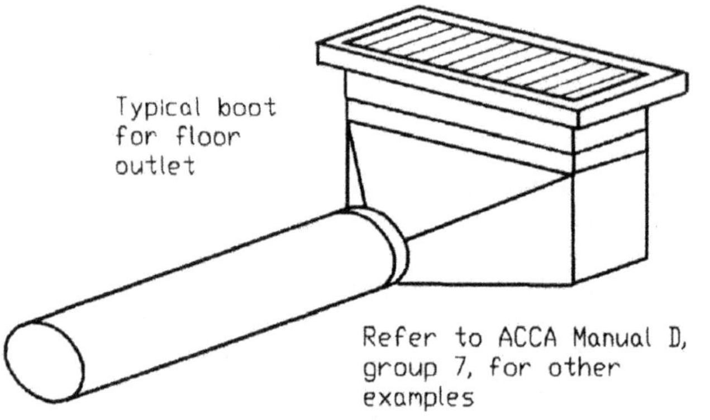

Floor and linear diffusers should never be attached flush with the wall of a supply trunk. The following figure shows the air flow pattern for a diffuser that has a short collar installed between the outlet and the supply duct. Note that in order to minimize "entrance related" distortions of the air pattern, an extractor is installed in the throat of the collar. The extractor helps turn the air toward the diffuser and this results in a more even distribution of air flow across the face of the outlet.

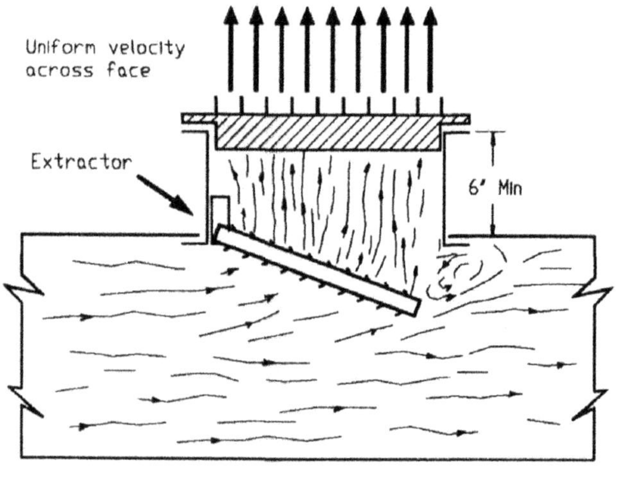

5-5 Total Air Pattern (Vertical Discharge Diffusers)

During heating, the total air pattern produced by vertical discharge diffusers will spread out and up the outside wall (A), across the ceiling (B), and part way down the inside wall (C). Also, the low, cooler air near the floor (D) is entrained and mixed quickly with the primary air which helps provide excellent heating performance.

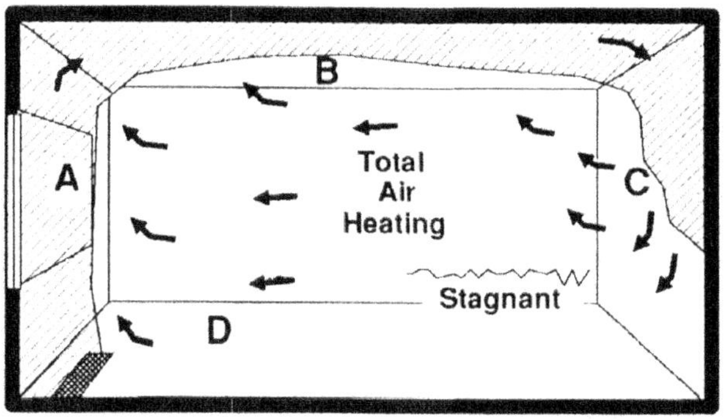

During cooling, the total air pattern of vertical discharge diffusers may or may not reach the ceiling level (A), depending on the throw. This is because of the tendency of the total air to fold back on the spreading primary air stream (B). The higher the total air can be projected, the better the cooling performance. Ideally the throw should reach the ceiling.

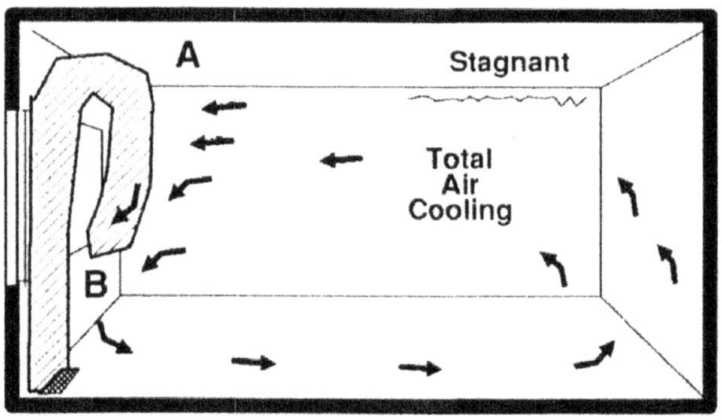

5-6 Total Air Pattern (Vertical Discharge Registers)

The total air pattern during heating produced by vertical discharge registers is similar to but not as wide as that provided by vertical discharge diffusers. A somewhat larger stagnant zone develops and the non-spreading primary air pattern does not counteract the descending natural convection air currents as well as the spreading pattern of the diffuser. Compare the total air pattern in the figure below to the heating air pattern for the vertical discharge diffuser on the previous page.

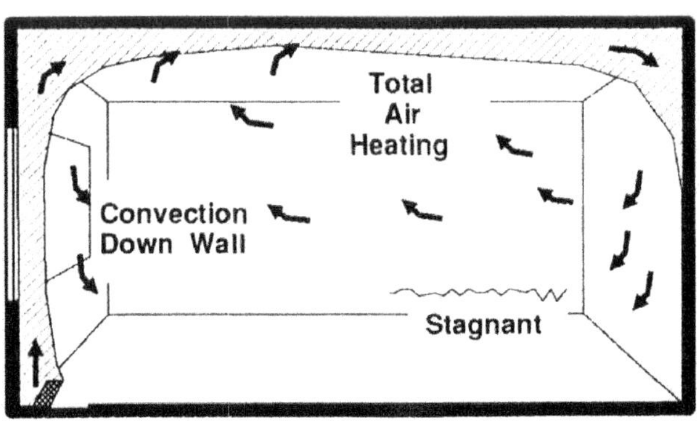

However, cooling performance is better with the non-spreading floor or baseboard register than with the diffuser. The total pattern is more likely to reach the ceiling and spread out over the ceiling, eventually dropping into the occupied zone (A). Compared with the total air pattern of the floor diffuser, the total air pattern of the register has less tendency to fold back on the primary jet.

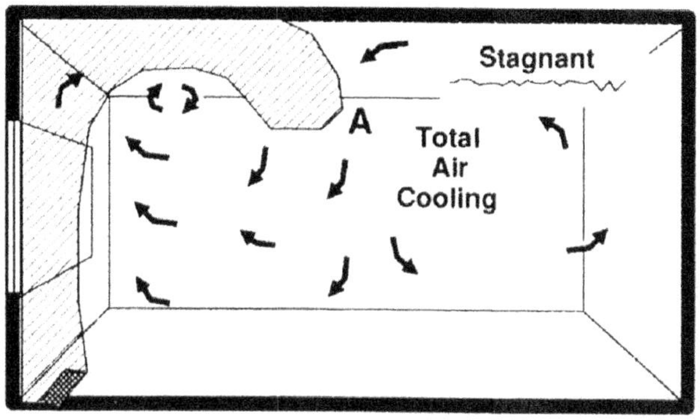

5-7 Total Air Pattern (Horizontal Discharge Outlets)

During heating, the total air pattern produced by these outlets will spread out and rise as it penetrates into the occupied zone. A good deal of secondary mixing takes place to the left (A) and right (B) of the total air pattern. Although the total air pattern appears to be acceptable for heating, complaints of drafts can be expected, because relatively high velocity air (in the primary air pattern) is projected directly into the occupied zone. To minimize the likelihood of high velocity air within the occupied zone, supply outlet velocities should be limited to 300 fpm or less.

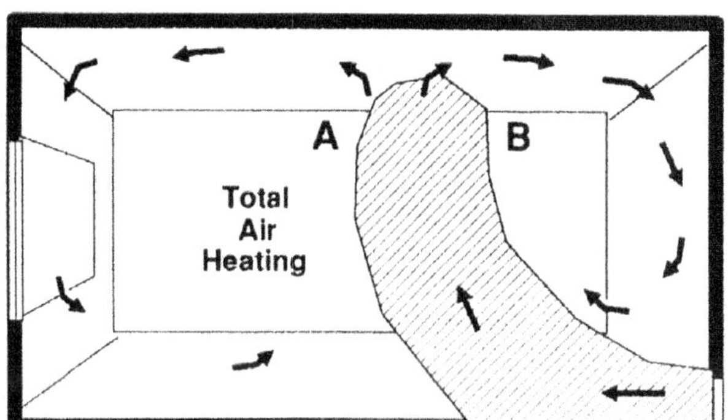

During cooling, the higher density of the supply air causes the total air pattern to just lay near the floor. Most of the occupied zone is located outside of the primary air pattern. Again, complaints of drafts might be expected because of the relatively high velocity air that is projected directly into the occupied zone. For this reason low sidewall horizontally discharging outlets are not recommended for cooling.

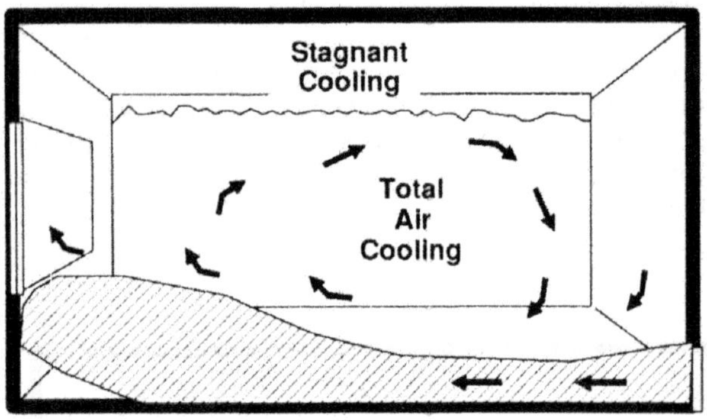

5-8 Basement Heating

One application for low sidewall horizontal discharge outlets could be in below grade or basement rec-rooms where cooling was not a particular concern, say in Northern climates. The diagrams below compare a low inside wall diffuser with a ceiling diffuser. With the low side wall diffuser the stagnant zone is close to the floor and the floor surface temperatures are generally comfortable except they are a little cool along the outside walls. With a ceiling diffuser the floor is generally colder and the conditions are noticable colder along the outside walls.

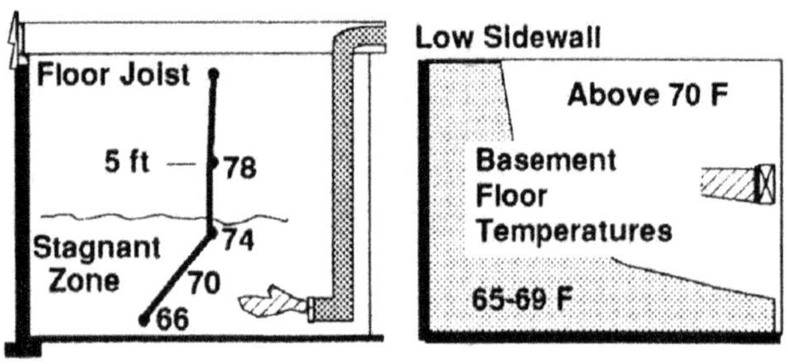

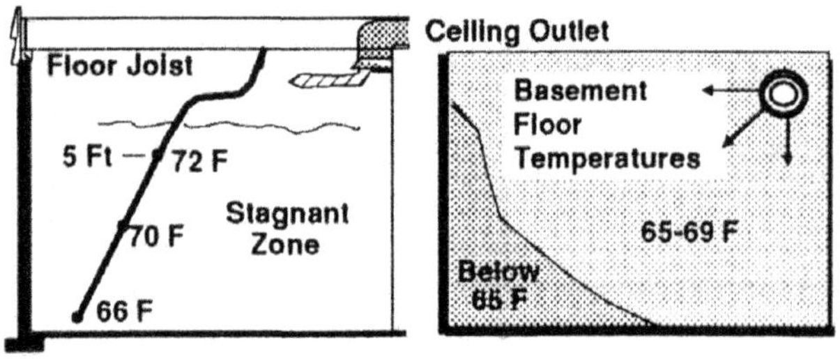

However, the low sidewall outlet is not the only choice for heating below grade spaces. Vertical discharge outlets located along the outside wall also perform quite well in this application. The diagrams below show that the stagnant zone that is associated with inside wall, horizontal discharge outlets, outside wall, and vertical discharge outlets is approximately the same. However, the outside wall outlet does seem to do a better job of floor warming. Note the much smaller shaded area indicating a larger area of warm floor temperature.

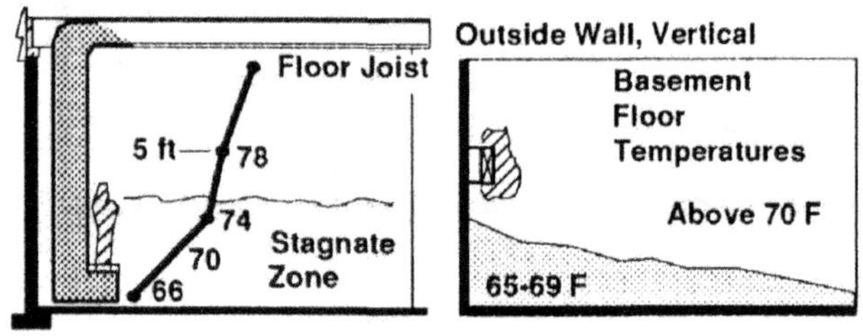

5-9 Stratification (Vertical Discharge Diffusers)

During heating a small stagnant zone does develop near the floor toward the opposite side of the room. However, this stagnate area is small and is located, for the most part, outside of the occupied zone.

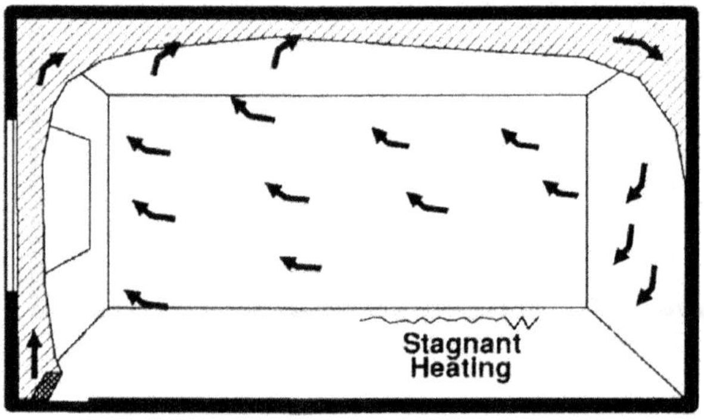

During cooling, a stagnate zone occurs near the ceiling at the opposite side of the room. The size of this stagnate zone depends on how effectively the diffuser throws the air toward the ceiling. In general, when the throw is equal to or a little more than the ceiling height, the stagnate zone does not cause objectionable conditions within the occupied zone.

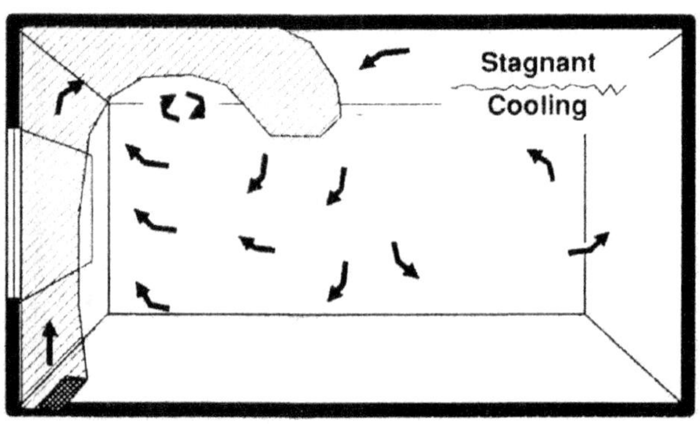

5-10 Stratification (Vertical Discharge Registers)

The stagnate zones produced by vertical discharge registers are similar to the stagnate zones produced by vertical discharge diffusers. The only difference is that during heating the stagnant zone is a little larger and during cooling the stagnant zone is a little smaller. Heating performance is reduced because the register has less spread than the diffuser and cooling performance is improved because the register has more throw than the diffuser.

5-11 Stratification (Horizontal Discharge Outlets)

During heating, horizontal discharge diffusers and registers do not produce a significant zone of stagnate air. The total air pattern and the secondary air patterns insure that the air within the occupied zone is in continuous motion.

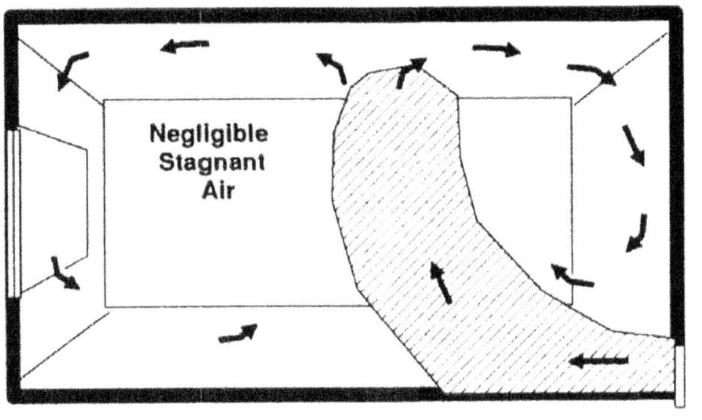

However, a large stagnate zone which extends completely across the room occurs during cooling. This zone begins at a little above the floor level and extends up through the occupied zone. Because the size of this stagnate zone is so large, horizontal discharge diffusers and registers are not recommended for cooling.

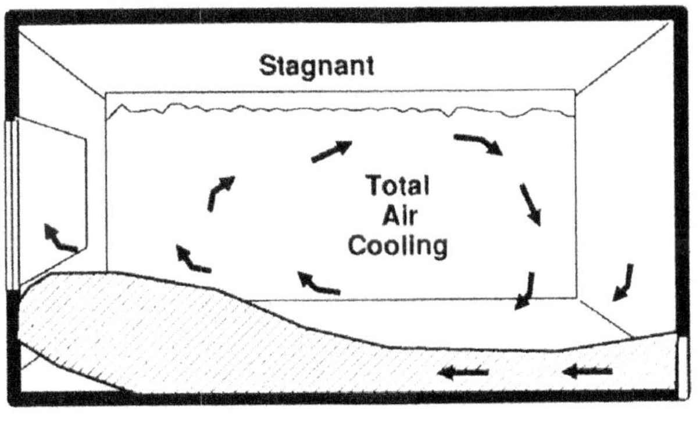

5-12 Floor and Low Sidewall Outlet Characteristics

The following comments are intended to illustrate some general performance characteristics of floor and low sidewall outlets. In actual practice the effects described below will vary depending on make, model, size and air flow rate.

Offset drafts caused by sliding glass doors:

Sliding glass doors are a source of cold drafts caused by strong descending natural convection currents and air leakage. Providing acceptable comfort at the floor level near sliding glass doors can be very difficult during very cold weather. While the performance of floor, low sidewall and extended baseboard diffusers are generally comparable, there is evidence that floor diffusers are better for heating rooms that have a sliding glass door.

The following figure below summarizes the results of a test that shows that low sidewall registers are not very effective at offsetting the cold convection currents that flow down the inside surface of sliding glass doors. In this test 210 CFM of supply air was delivered by a 6 x 10 sidewall diffuser. (The load calculation indicated that only 100 CFM was required to heat the room.) Even though the supply CFM was increased to more than twice the design CFM the temperature at the 4-inch level was only 66 degrees. This is unacceptable because the ASHRAE comfort standard states that the temperature at the 4-inch level should be within 5 degrees of the temperature at the 67-inch level. In other words the temperature at the 4-inch level should have been 70 F or higher to meet the ASHRAE comfort standard.

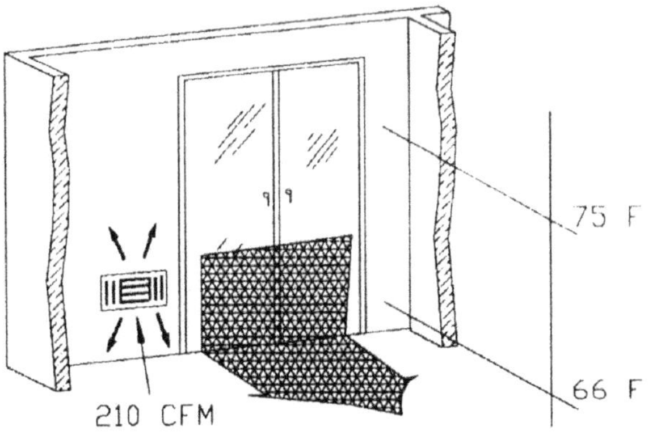

In another test a 4 x 10 floor diffuser with curved blades was inserted in front of the glass doors. In this case only 80 CFM, (less than the required heating CFM) was supplied. Note that with the floor diffuser the temperature at the 4-inch level was improved by 5 degrees (66 F to 71 F) compared to the sidewall diffuser which delivered 210 CFM.

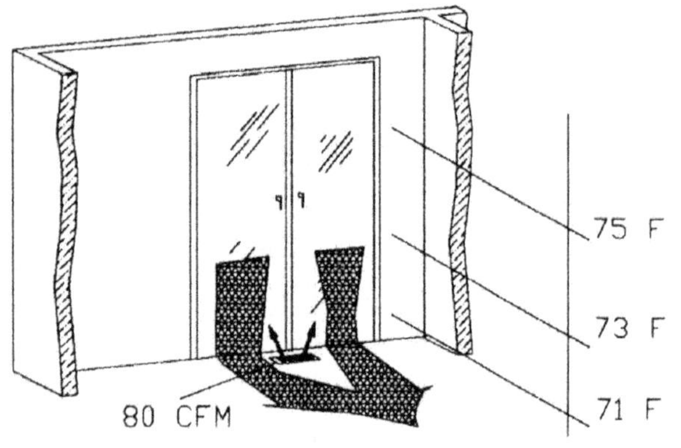

In a final test, two 2-1/4 x 10 standard floor diffusers were installed and adjusted for a combined air flow rate of 96 CFM. In this case the floor level temperature was increased to 71 F. This is comparable to the performance of the single, curved blade diffuser operating at 80 CFM.

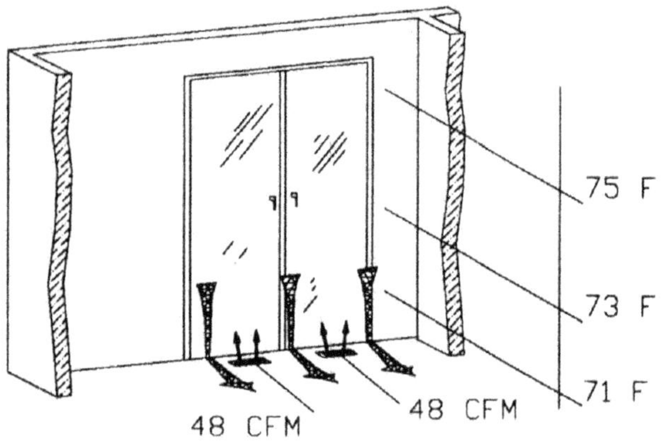

The obvious conclusion is that when sliding glass doors are involved, installation of floor diffuser directly in front of the door is by far superior to installing a low sidewall diffuser next to the door.

Effect of drapes in front of sliding glass doors:

Tests were conducted (during heating) with the a floor diffuser located between full length drapes and a sliding glass door (Sketch A) and then with the floor diffuser located in front of the full length drapes (Sketch B). The shaded areas in the Sketches A and B show regions of cold temperatures at the ankle level, with the darker areas denoting the coldest temperatures. Note there is substantially more floor level area below room temperature when the drapes are placed in front of the diffusers Sketch A). In this position, the drapes prevent the induction of floor level air by the primary jet of the diffusers.

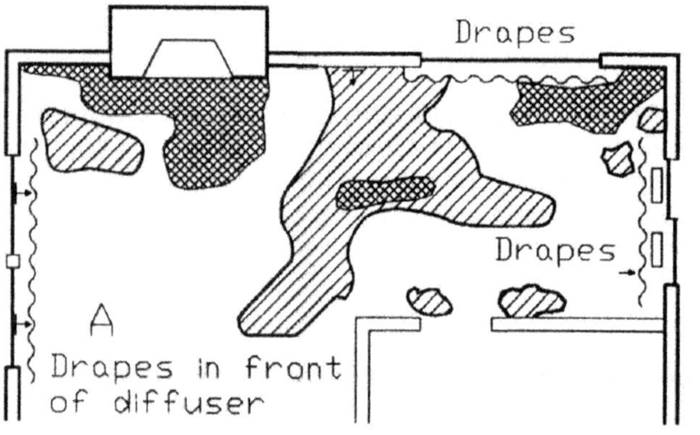

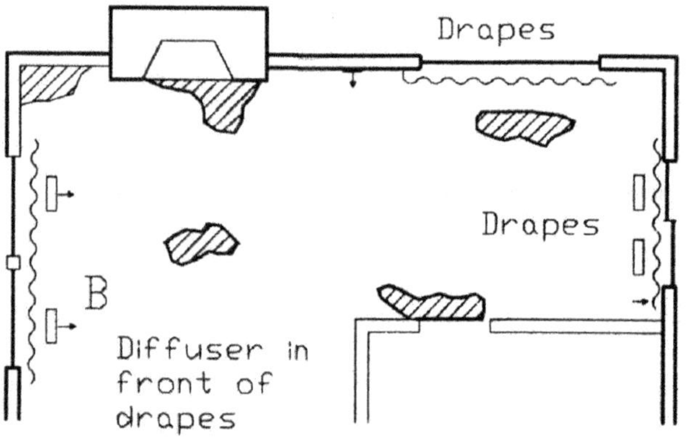

Ability to offset drafts created by windows:

It is important to place a diffuser directly below a window so it can intercept the cold convection currents cascading down from the window. In the diagram below, the performance of a diffuser which is centered under the window is compared with the performance of a diffuser that is off-set (by 4 feet) from the centerline of the window. Note that the ankle level temperatures in the room decreased from 68 F to 66 F when the outlet was installed off center.

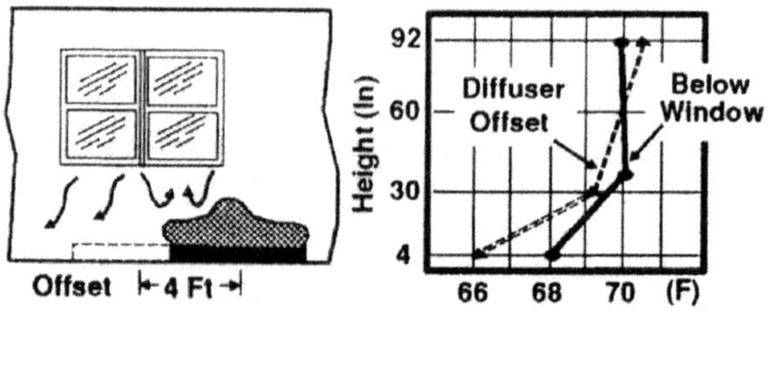

Window drapes affect cooling performance:

Full length draperies can have a significant affect on the cooling performance of diffusers which are installed below windows. The diagrams below show that when the drapes are open, the stagnant region extends down from the ceiling to about 30 inches above the floor and when the drapes are closed the stagnant region is reduced, only reaching within 60 inches of the floor. This is because with the drapes closed, they acted as a duct to project the primary air higher into the room thus improving performance.

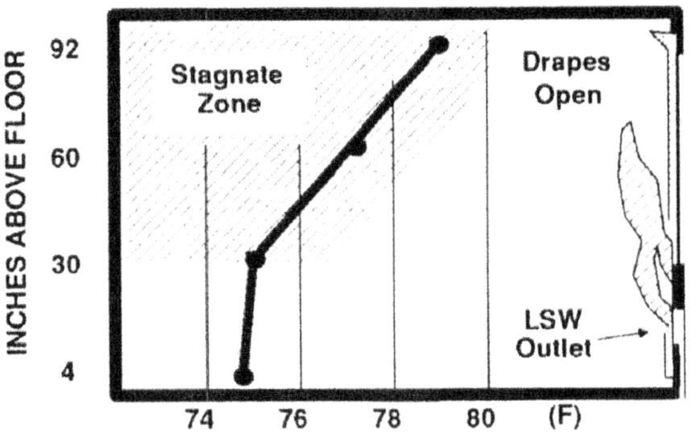

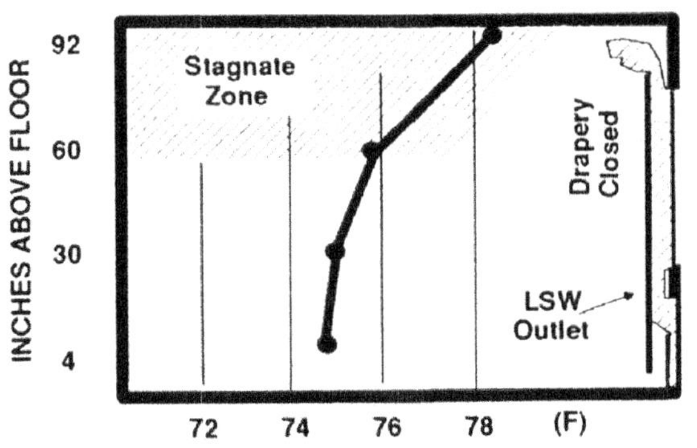

During cooling short drapes can also be a problem. In the following diagrams, an extended baseboard diffuser is positioned under a window covered by short drapes. With the drapes closed they tend to "puff out" from the window and deflect the primary jet downward. This results in a stagnant region extending almost the full height of the room.

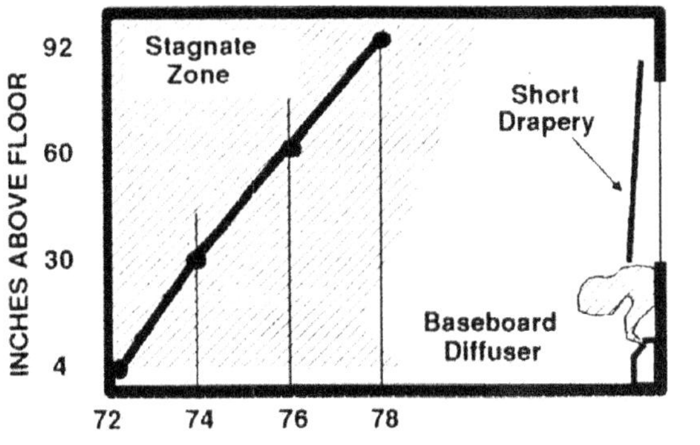

However, if a retainer is used to keep the drapes tight against the window ledge the stagnate zone only reaches to within 60 inches of the floor. Restraining the drapes allowed the outlet to project the primary air high into the room, reducing the stagnant region.

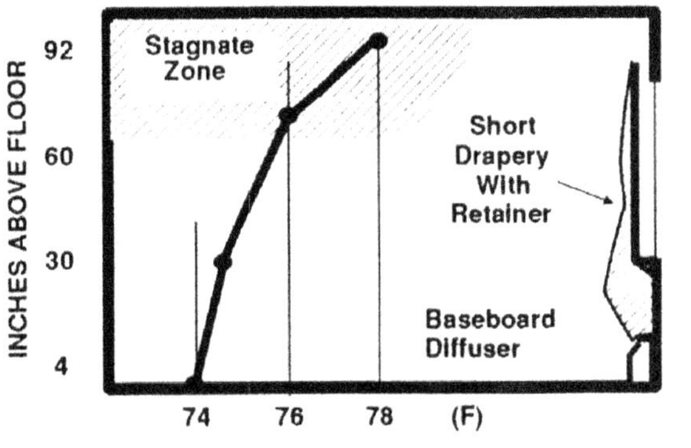

5-13 Location Of Returns

In general the location of the return has no effect on the air pattern within the occupied zone and negligible effect on the shape and size of a stagnate zone.

Section 6

Pressure Loss Characteristics of Supply Outlets

As with any component that is inserted into a duct system, there is a pressure loss associated with moving air through a supply outlet. This pressure loss is a consequence of friction, turbulence and velocity changes. Before discussing the nature of outlet losses, a brief review of the basic principles which govern the pressure distribution in the duct system is necessary.

6-1 Inches of Water

It doesn't take much pressure to raise a column of water an inch or two. This sensitivity makes the water gage an ideal instrument for measuring the relatively small pressures that are associated with fans and duct systems. Therefore the terms "Inches, Water Gage" or "Inches, Water Column" are almost always used to describe air-side pressure measurements.

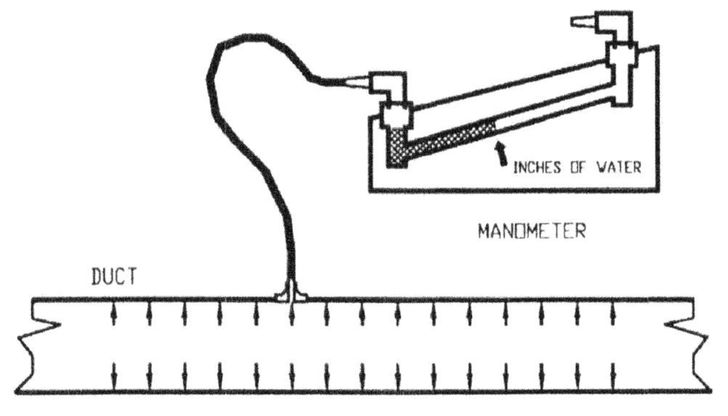

6-2 Static Pressure

The figure below illustrates how static pressure is measured. In this test a hole is punched in the wall of a duct and connected, by a tube, to one side of an inclined manometer (pressure gage). As the connection is made we would see the water level in the manometer rise. This rise in water level indicates that the pressure inside of the duct is higher than the pressure outside of the duct. In this test the manometer measures the pressure that is tending to "burst" the duct. (Note that the direction of the air flow is at right angles to the static pressure tap.)

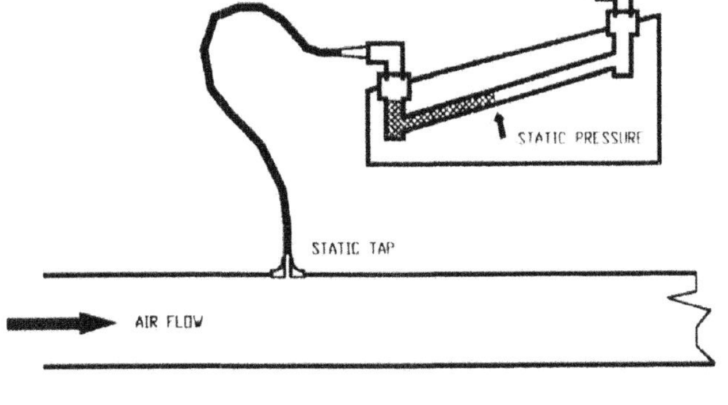

6-3 Total Pressure

The figure below illustrates how total pressure is measured. In this test an L-shaped tube is inserted through the wall of the duct (with the open end facing the flow) and connected to one side of the manometer. Once again we would observe water rising in the manometer, but this time it would rise to a greater height. In this test the L-shaped tube still measures the static pressure and it also measures the additional pressure that is created when the moving air flow is brought to a stop by the open end of the tube.

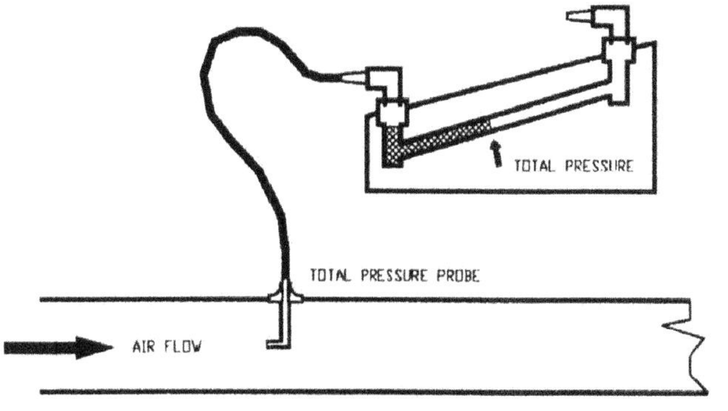

6-4 Velocity Pressure

The velocity pressure is equal to the difference between the total pressure and the static pressure. In the figures above it is equal to the difference in the height of the two water columns. The figure below illustrates how to measure velocity pressure directly. In this test the pressure taps are connected to the manometer so that the total pressure tends to increase the water level while the static pressure tends to decrease the water level. In effect, this subtracts the static pressure from the total pressure and the resulting rise in water level is exactly equal to the velocity pressure.

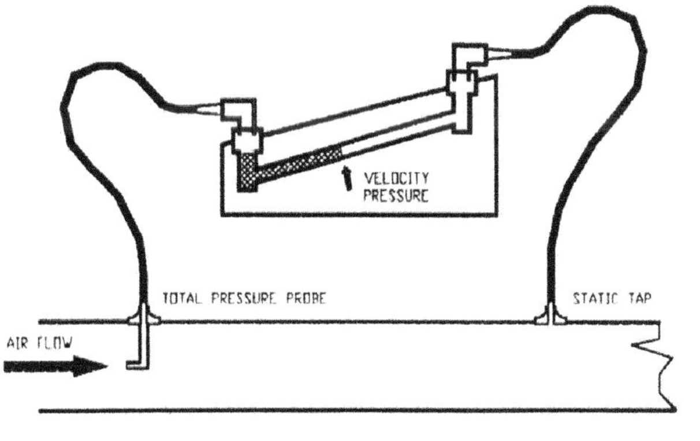

Velocity pressure depends on the velocity of the air stream. Faster moving air has a higher velocity pressure than slower moving air. Also note that velocity can be converted to static pressure and visa-versa.

6-5 Static Regain

As discussed above, when air moves through a duct the total pressure (sum of the velocity and static pressure) always decreases in the direction of air flow. However, the static pressure can increase or decrease depending upon the changes in the duct size or, when air is extracted at a takeoff, the flow rate.

When the velocity of a moving air stream is decreased most of the corresponding decrease in velocity pressure is converted to static pressure. The term "static regain" is used to describe the conversion of velocity pressure into static pressure. Conversely, when the velocity of a moving air stream is increased it does so at the expense of the static pressure.

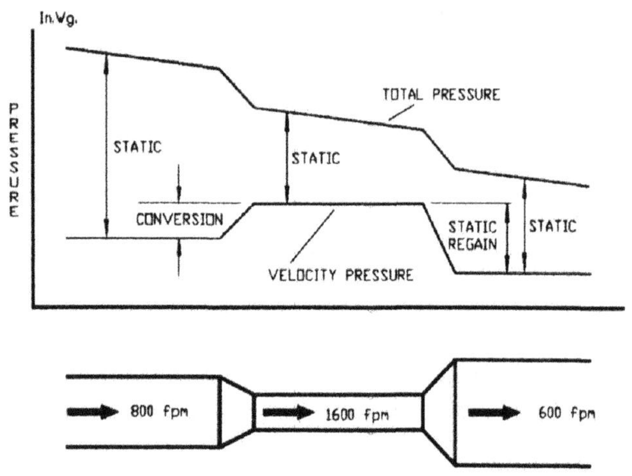

Note that an abrupt expansion is very inefficient; little or no regain occurs and a high pressure loss occurs. By gradually transitioning from one duct size to the other an efficient conversion of velocity pressure to static pressure is achieved resulting in more static regain and less pressure loss.

6-6 Pressure Loss with Constant Velocity

To determine the net pressure loss for a section of ductwork of uniform size, we can simply observe the difference between the static pressure at two locations.

In the figure below the static pressure at point B is less than the static pressure at point A. This loss of static pressure indicates that friction has had an effect on the air flow. In this case the velocity (and therefore, the velocity pressure) does not change because the duct size is constant. Since the velocity pressure is constant, the loss in static pressure must be exactly equal to the loss in total pressure. Thus the loss of static pressure must be exactly equal to the net pressure loss which is equal to the loss that is associated with overcoming friction as the air flows from point A to point B.

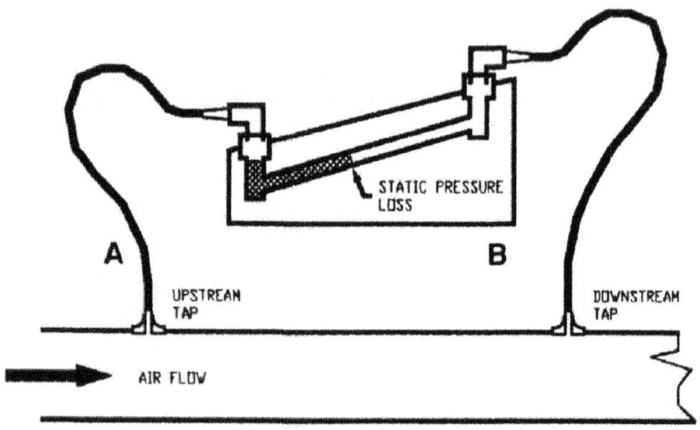

6-7 Pressure Loss with Changing Velocity

To determine the net pressure loss for a section of ductwork of changing size, we must observe the difference between the total pressure at two locations.

Refer to the figure below and note that the total pressure at B is less than the total pressure at A. Once again, a loss of total pressure indicates that friction has had an effect on the air flow. In this case however, the velocity (and therefore the velocity pressure) does change because the duct size has increased. Consequently the change in static pressure is not equal to the net pressure loss or to the friction loss. (Note that in this example the static pressure actually could increase if the friction loss is less than the static regain.) When there is a change in velocity of the moving air stream we must measure the change in total pressure between the two locations to determine the net pressure loss.

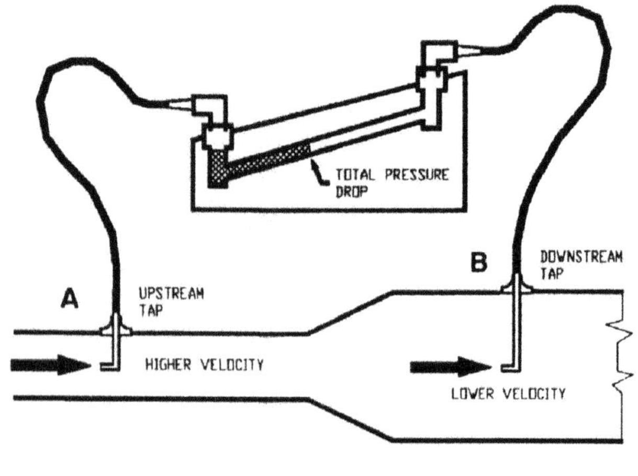

6-8 Pressure Loss of Supply Outlets

Manufacturers' catalogs usually report on the pressure loss that is associated with supply outlets. Inspection of various catalogs indicates that this pressure loss is either reported as a static pressure loss or as a total pressure loss. This inconsistency is annoying but it is not of significant importance. (For face velocities of 600 fpm or less the difference between the total pressure and the static pressure values will not exceed 0.03 inches of water.) The diagram below shows how the pressure drop across the outlet is measured.

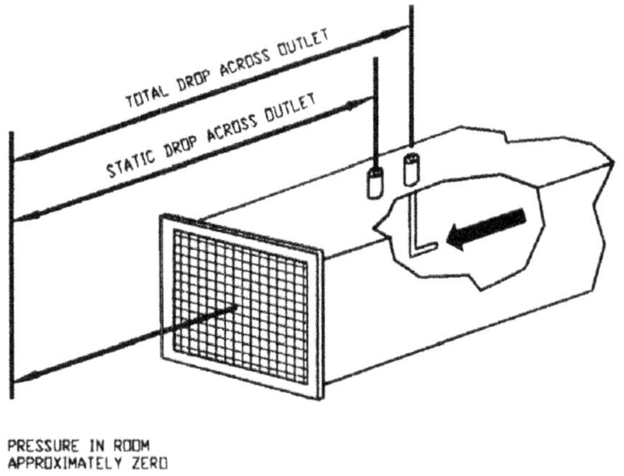

- The static pressure loss is equal to the difference between the static pressure upstream (ahead) of the outlet, minus the static pressure in the test room. Because the static pressure in the test room is approximately zero, the static pressure loss of the outlet is equal to the upstream static pressure.

- The total pressure loss is equal to the difference between the total pressure upstream of the outlet, minus the total pressure in the test room. Note that after the air moves away from the outlet the velocity is very low and the velocity pressure is negligible. Since the static pressure and the velocity pressure in the room are approximately zero the total pressure in the test room is also approximately zero. Therefore, the total pressure drop across the outlet is simply equal to value of the total upstream pressure.

6-9 Accessory Pressure Loss

Because supply outlets are often equipped with dampers and/or connected to a fitting (boot, stackhead, etc.) it is reasonable to measure and report the pressure loss of the fitting-outlet combination. This method of reporting outlet pressure loss is sometimes used for residential and light commercial registers, diffusers and grilles. Before using catalog pressure loss data it is important to verify whether or not the pressure loss data includes a damper and/or a connecting fitting.

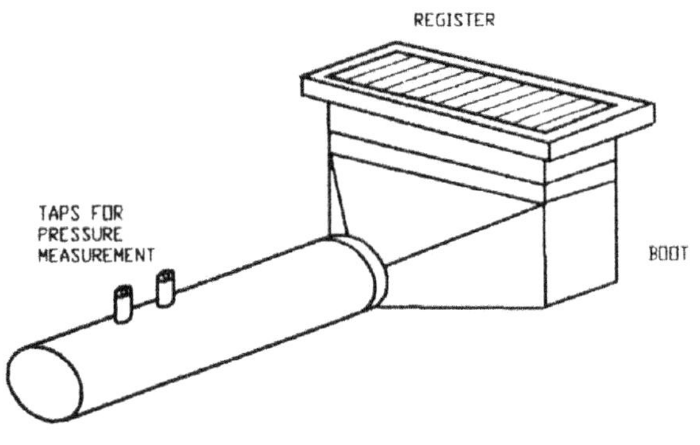

Also note that the total pressure required to "push" air through the supply outlet into the room is not a constant but varies with CFM flowing through the outlet. The graph below provides one example of how the outlet pressure loss varies with the flow rate. Note that if the CFM is doubled from 60 CFM to 120 CFM, the pressure drop increases from 0.024 inches w.g. to 0.098 inches w.g., or by a factor of 4! Thus, the pressure drop varies approximately as the square of the flow rate.

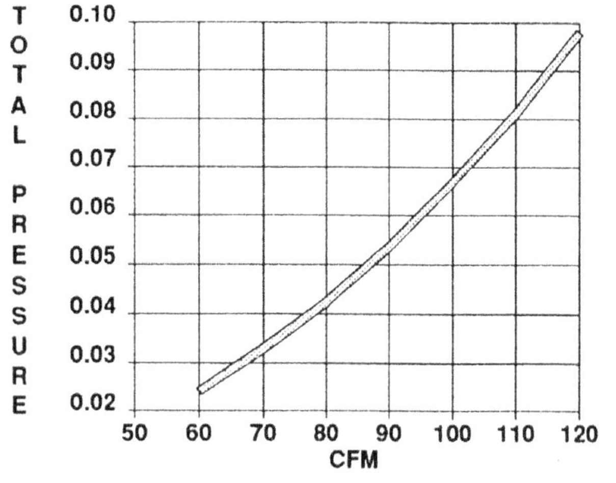

6-10 Optimum Outlet-Fitting Geometry

When the pressure loss of the floor diffuser and the fitting are measured together, there is an optimum combination of branch feeder duct size and outlet size which offers the least resistance to flow. This effect is illustrated in the diagram below. Note that the 4 x 10 diffuser works best with the 5-inch round runout. However, the additional pressure loss that is associated with the less efficient configurations can add approximately 20 feet of equivalent length to the duct run.

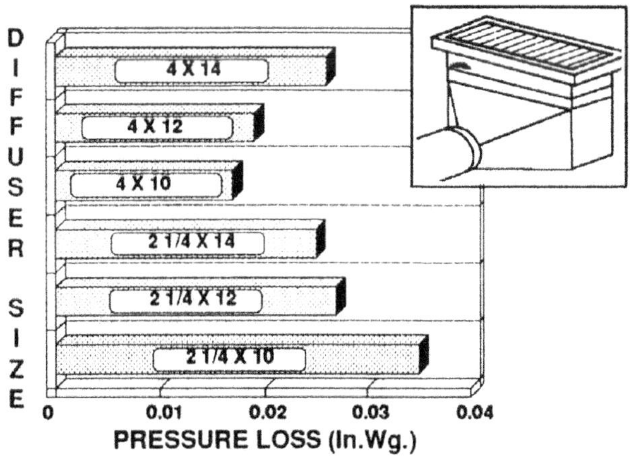

The following table summarizes the various combinations of floor diffuser sizes and branch run-out sizes which will result in minimal pressure loss.

	Runout Size			
Outlet Size	3	4	5	6
2 x 10	x	x		
2 x 12	x	x		
2 x 14	x	x	x	
4 x 10	x	x	x	
4 x 12	x	x	x	x
6 x 10		x	x	x
6 x 12		x	x	x
6 x 14			x	x

There are also optimum branch runout to wall stack combinations which minimize the pressure loss associated with high sidewall registers. Table below indicates the optimum run out diameter for various standard wall stack sizes. Note that this table only applies to high sidewall applications. (The stack/runout relationship is not a contributing factor for the low sidewall configuration because the supply air is unable to completely fill the stack before it is discharging through the register.)

	Runout Size			
Stack Size	5	6	7	8
2 x 10	x			
2 x 12		x		
2 x 14			x	
3 x 10		x		
3 x 12			x	
3 x 14				x

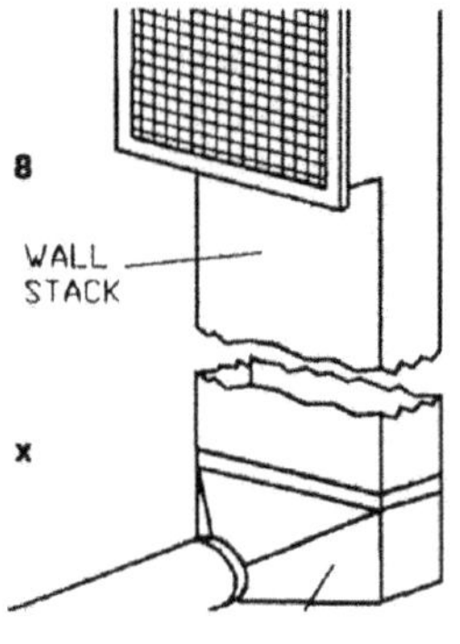

The pressure loss through a ceiling diffuser can be reduced by adding a collar of adequate length. In the illustration below, the pressure loss through a 12-inch round ceiling diffuser mounted at the bottom of a rectangular duct is shown for three different lengths of connecting collar. Note that the addition of a 4-inch short collar reduced the pressure loss from 0.35 to 0.28 inches w.g.. Also note that as the length of the collar was increased the pressure loss increased, from 0.28 with a 4-inch collar to 0.30 with a 14-inch collar. Additional tests showed that after 20 inches of collar was added the pressure loss was the same as the loss that would occur with no collar.

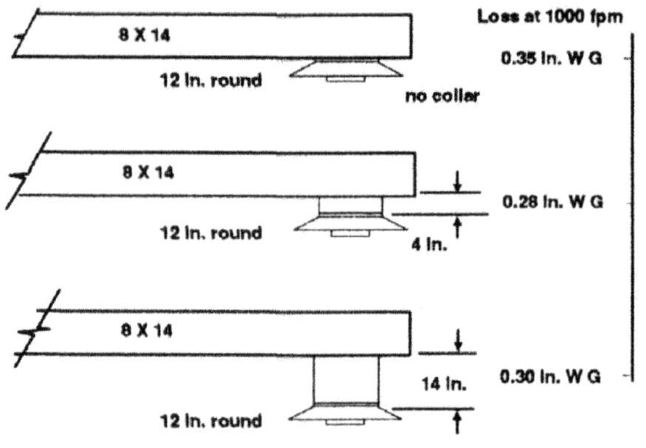

The "collar effect" also applies to low sidewall outlets. In general the pressure loss for a boot/stackhead/register combination is less when 1 to 2 feet of wall stack is inserted between the stackhead and transition boot. In the figure below (for the same air flow rate, same grille size, etc.) combination (B) would experience a smaller pressure loss than combination (A).

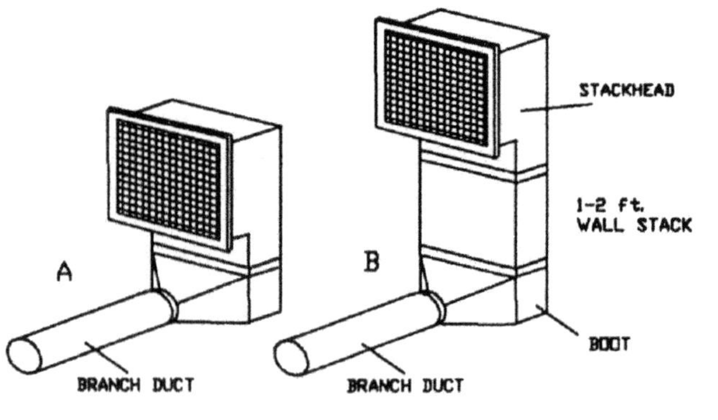

Section 7

Performance of Return Inlets

For every cubic foot of air that is supplied by the conditioning equipment, an equal amount of air must be returned to the equipment. An adequate return system is absolutely necessary or the performance of the cooling and/or heating system, particularly the equipment and the supply outlets, will be unacceptable.

7-1 Types of Return Inlets

Various types of return grilles and registers are designed and styled to cover wall, floor and ceiling openings while others are designed for installation at baseboard locations. (Grilles and registers are similar in appearance and performance except registers are equipped with a damper that can be used to control the amount of air flowing through the return inlet.) Some returns are purposely designed to match the style of the supply outlet. For example, an extended baseboard supply outlet may have a matching return inlet, or, in commercial applications, a ceiling return may be similar in appearance to the supply diffuser. Also note that any supply outlet, including ceiling diffusers and slot diffusers, could also be used as a return inlet.

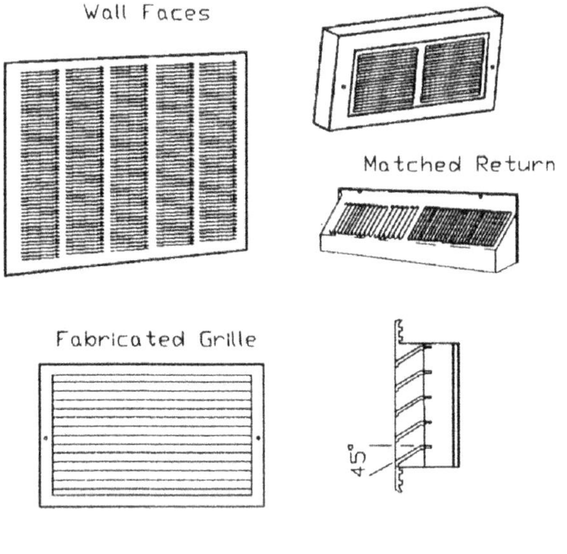

In addition, there are a number of types of return grilles and registers that are especially designed to solve certain architectural problems.

- Door grilles provide a return path from a room to a corridor which functions as a return air trunk duct.

- Vision proof, sound retarding grilles provide privacy when air from a room which does not have a return intake is returned through an opening in a partition or door panel.

- Combination ceiling supply/return units reduce the total number of supply outlets and return inlets which must be purchased when returns are required in every room. Generally, the return portion is in the center section with the perimeter used as a supply. Often, the return air flow is limited to about 90 percent of the supply CFM to avoid short circuiting of the supply into the return portion of the unit.

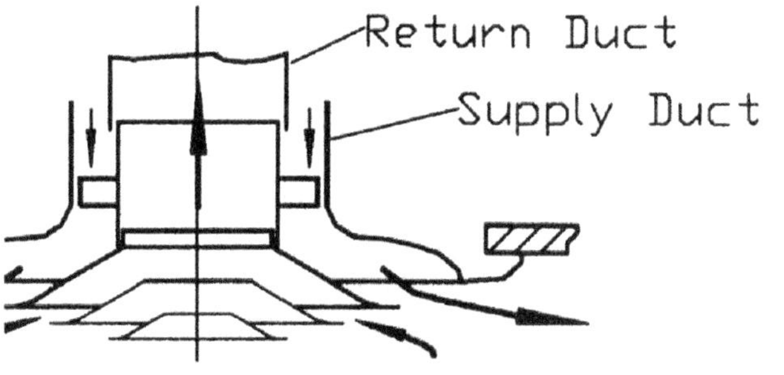

- Filter grilles can be used to make filter changing more convenient when the equipment location makes filter service difficult.

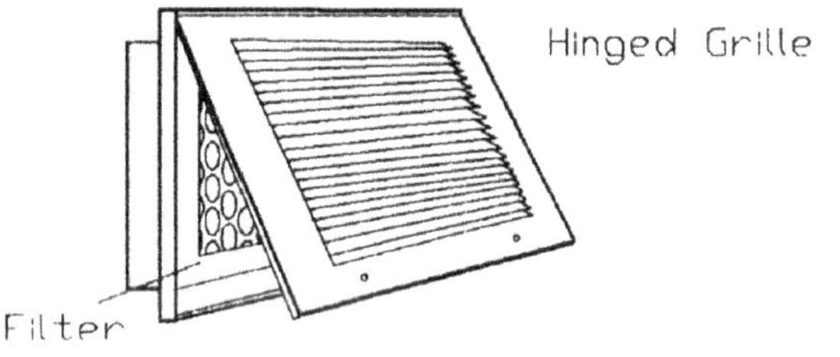

7-2 Negligible Effect on Room Air Patterns

Laboratory tests show that the influence that a return intake has on room air patterns is limited to a tiny area within a few feet of the intake. The figure below illustrates this point; note that even with an unreasonably high face velocity of 1000 fpm the velocity of the room air, just inches away, is only 50 fpm.

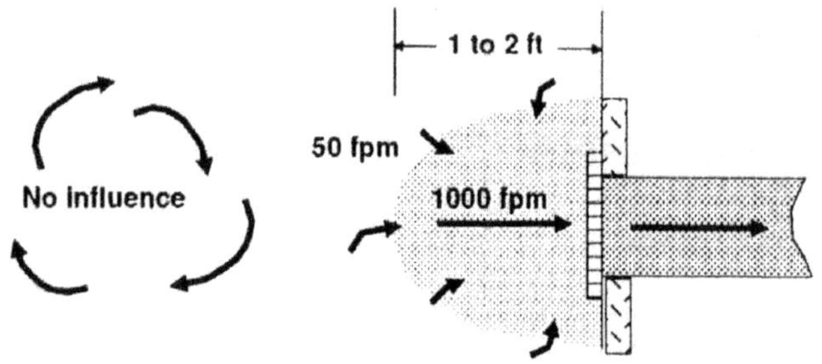

Another example of the meager "pulling power" of return inlets is provided by the following graph. This graph consists of a series of curves which show how the room air motion is affected by the face velocity at the return grille. Note that the face velocity at the inlet has a negligible effect on the motion of the surrounding air. (At about 15 inches, the air velocity is around 50 fpm and even with a grille face velocity of 1,000 fpm, the velocity is 50 fpm at 24 inches from the grille.)

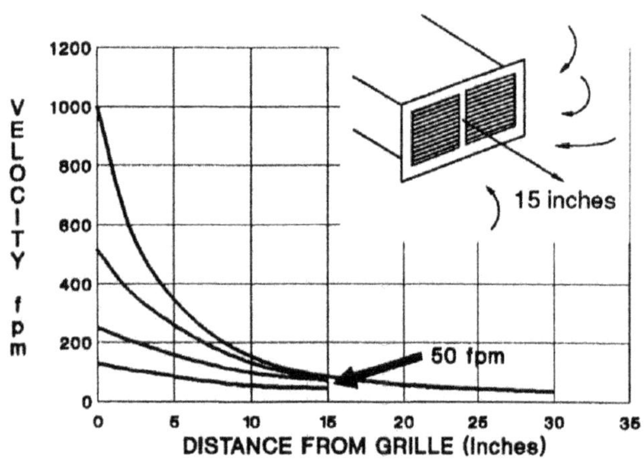

The obvious conclusion is that other disturbances to the room air motion patterns can exert a greater influence on room air motion than the return. Relocating returns cannot cure poor air distribution problems. Changes to the supply system can be far more effective in improving poor air circulation.

7-3 Guidelines for Locating Returns

Even though the influence of a return over air motion within a room is negligible, there are preferred return locations. As a general rule, the return should be located inside the stagnant region. The stagnant zone is characterized as an area that is not under the influence of the supply outlet but which is subject to natural convection currents.

- For "heating only" applications, the stagnant zone is at the floor level. This means the return should be placed at the floor level. This recommendation applies whether the supply outlets are in the floor or high in the sidewall or ceiling. In this arrangement, the coldest air is removed first and returned to the conditioning equipment.

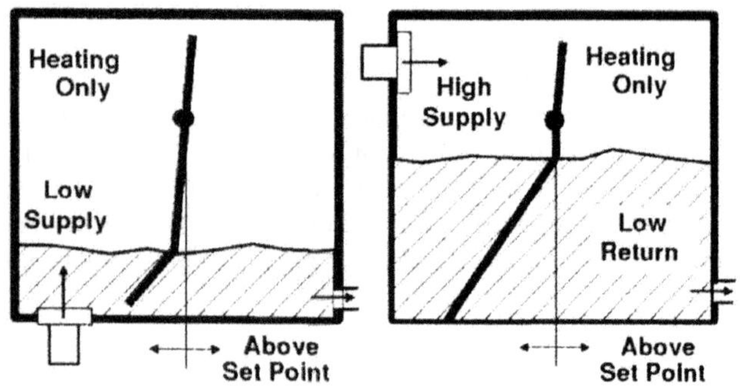

- For "cooling only" applications, the stagnant zone is near the ceiling. This means the returns should be placed high in the room. Again it does not matter if the supply outlets are high or low. In this arrangement, the warmest air is removed first and returned to the conditioning equipment.

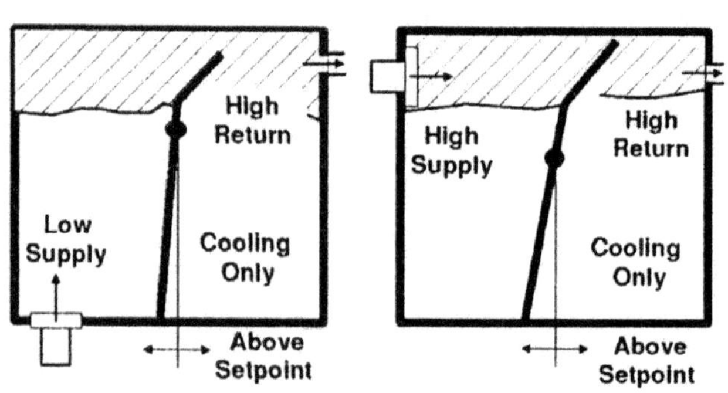

- In the case of a combination heating and cooling system, the return location depends on when the largest stagnant zone is likely to occur, during heating or during cooling. When perimeter supply outlets are placed low in the room, the largest stagnant zone will occur near the ceiling, during cooling. Therefore when low outlets are used the returns should be placed high in the room. When high supply outlets are used the heating performance is poor and the largest stagnant zone will occur near the floor, during heating. In this case the returns should be placed low in the room.

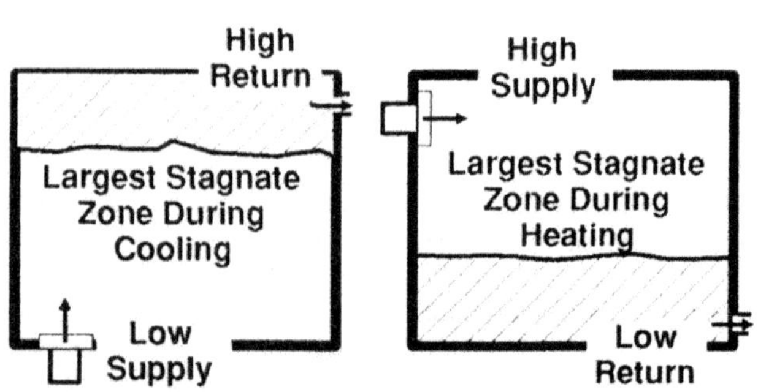

Section 8

Pressure Loss Characteristics of Return Inlets

As with supply outlets, there is pressure loss associated with drawing air through return inlet.

8-1 Inlet Pressure Loss

Inlet pressure losses published in manufacturers' catalogs. The inlet loss may be expressed as either a static or a total pressure drop. In measuring the the static pressure loss, a grille is placed on the end of a duct of the same size and the static pressure (SP) is measured far down stream. Because the pressure on the return side of the fan is below the (atmospheric) pressure in the test room, the return static pressure is measured as a negative pressure. Note that the total pressure is equal to the sum of the positive velocity pressure and, in this case, the negative static pressure.

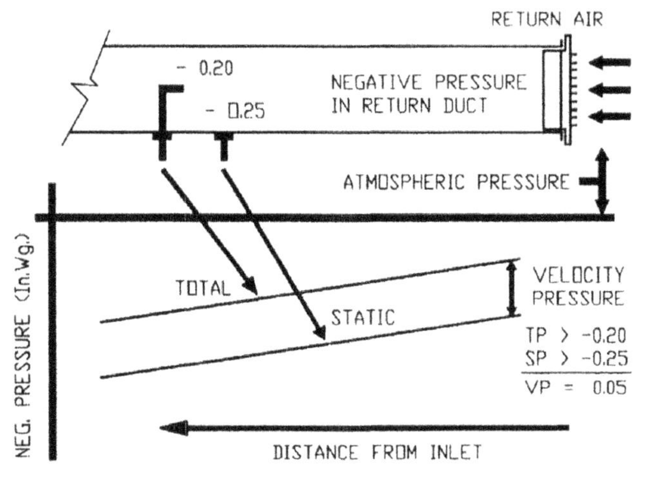

8-2 Pressure Loss is Affected by Inlet Geometry

The following series of diagrams and sketches illustrate how the entrance conditions affect the inlet flow patterns and pressure loses.

The following graph relates the grille free area to a loss coefficient "C" which represents the magnitude of the entrance pressure loss. Note that if the free area is decreased from 1.0 to 0.50 (a 50 percent decrease) the "straight in" loss increases from 1 to 3 (a 200 percent increase). This graph clearly demonstrates why the return grille size should be based on its free area rather than its rough overall dimensions.

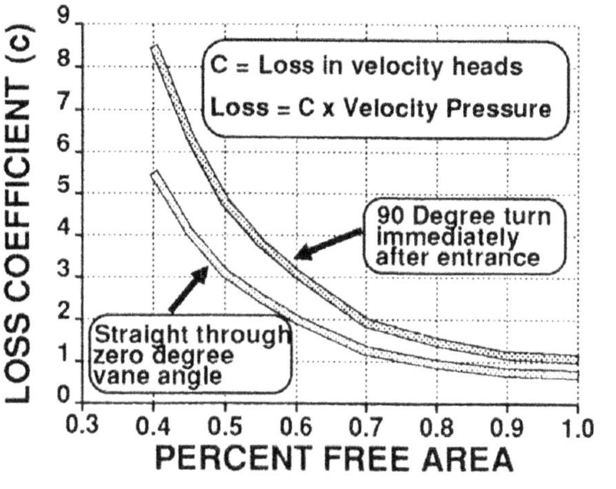

The sketch below illustrates how air contracts as it enters a duct that has no inlet grille. The lines which are squeezed together indicate a rapid acceleration of the stream. Note that the air stream does not immediately fill the duct, in fact it has a tendency to neck down. This contraction tends to increase the acceleration and the pressure drop. If a return grille was installed at the entrance, the pressure drop would be even greater because the free area would be reduced causing the velocities and corresponding accelerations to be even higher.

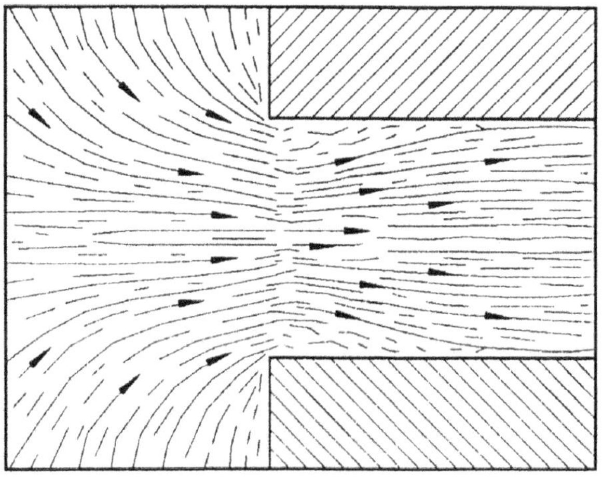

Designs which force the air to turn as soon as it passes through inlet opening are especially inefficient. For example, if the a return grille is installed in a vertical riser the air is forced to make a 90-degree turn as well as accelerate through the grille.

In this case the pressure loss of the grille could be increased by as much as 65 percent (as indicated by the graph in Section 8-2). To compensate for the turn, it is acceptable to select a grille with a free area that is larger than the free area of the return duct.

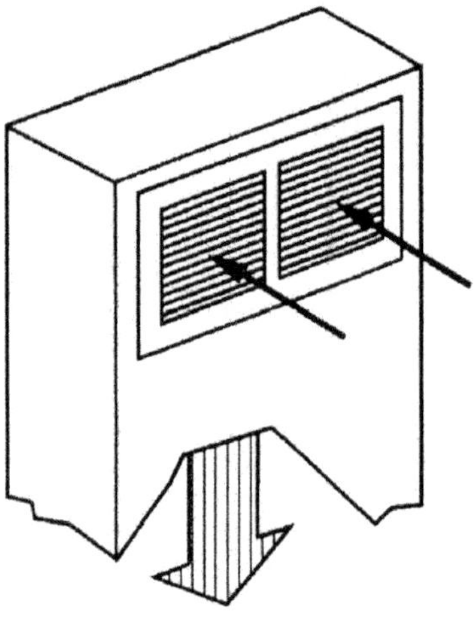

#116

Conversely, proper inlet vane adjustment can reduce inlet pressure losses. The following sketch illustrates how vanes of a grille that is installed in a stack head can be used to help turn the air through the 90-degree bend. Note how the main stream of high velocity air tends to bend into the riser and completly fill the duct. When the horizontal vanes are properly adjusted the flow pattern is improved and the inlet pressure loss is reduced.

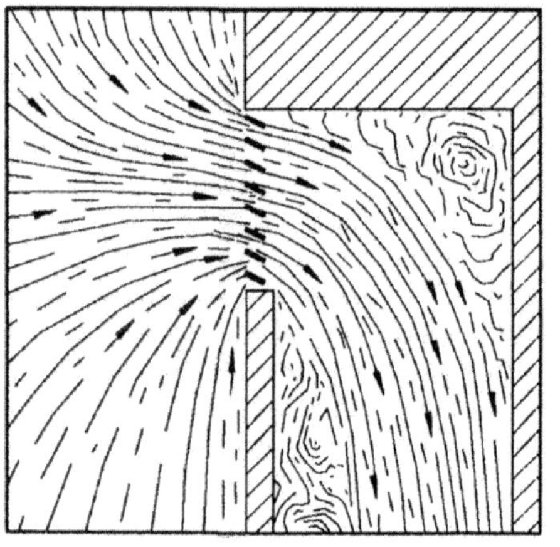

Improper inlet vane adjustments also affect inlet pressure losses. The sketch below illustrates what happens when adjustable vanes are set at a 45-degree angle to the direction of flow. Note that the eddy currents (random or circular air movement) and contractions of the stream inside the duct are quite unsymetrical and disorderly. Laboratory tests have shown that improperly adjusted inlet vanes have an adverse effect on the inlet pressure loss.

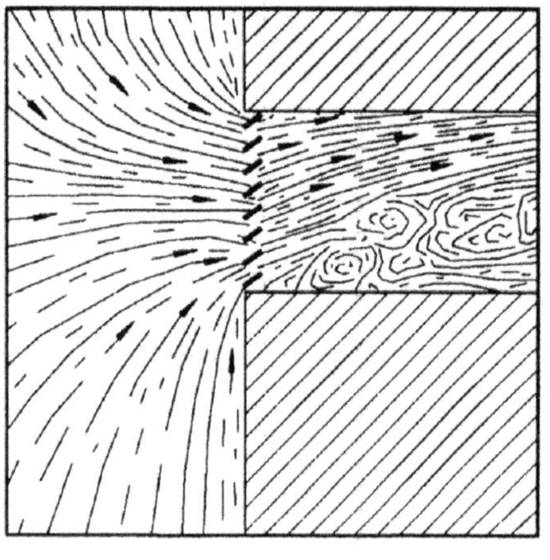

In the sketch below, the vanes of the stackhead grille have been readjusted to direct air up into the stackhead. Compare this sketch with the previous sketch. Note that the main stream now fills less of the duct, indicating a less efficient flow pattern and a higher inlet pressure loss. In some cases improper vane settings can result in a pressure loss as much as 200 percent higher than normal. In general, inlet vanes should be positioned to help turn the air and minimize pressure loss.

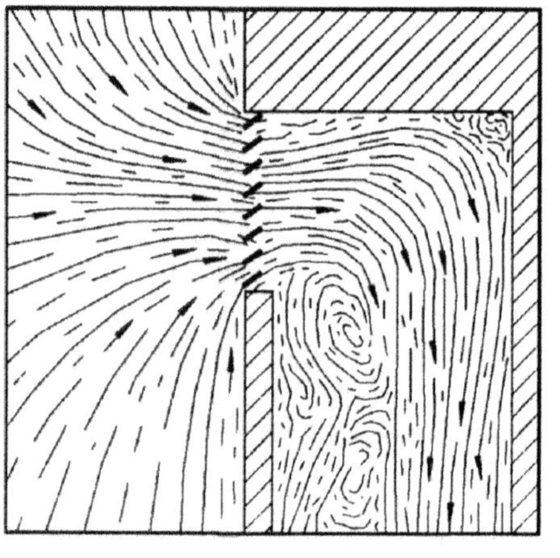

Section 9

Sound Considerations

Some manufacturers of registers, grilles and diffusers provide information on the sound characteristics of their devices. Others suggest minimum velocities to avoid noise problems. This section provides a review of the basic sound principles, which when properly applied, will help avoid noisy systems.

9-1 Frequency

Sounds are produced by disturbances in the surrounding air. These disturbances may be caused by a vibration, a single impact or a series of impacts. Once a sound is produced it exists as a pressure wave which moves through the air (propagates) in a manner that is similar to the way a wave moves through water. Consider this analogy.

A stone thrown into the water creates ripples (waves) which spread out (propagate) in all directions from the point of impact. The size of the ripples depends upon the size of the stone and how fast it was moving when it hit the water. An observer stationed at some distance from the point of impact would observe a series of waves moving by. The space between succeeding ripples is called the wave length. The passing of one complete wave is called a cycle. The number of waves (cycles) that pass by the observer in a given amount of time, say one second, is called the frequency. Frequency is commonly measured in cycles per second (cps).

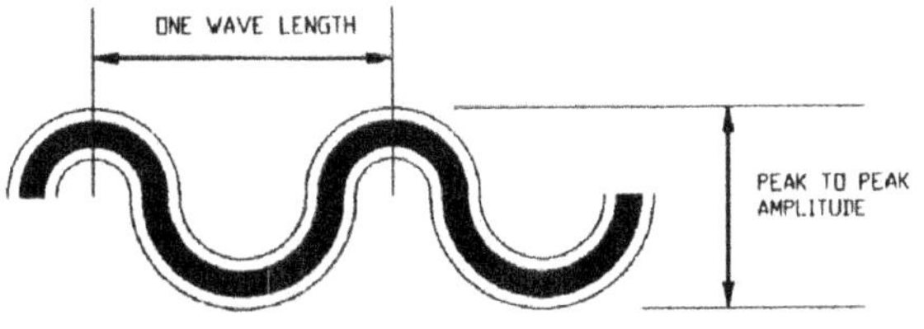

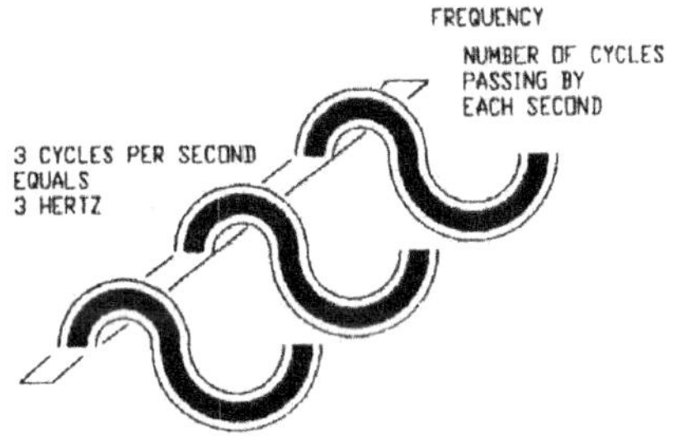

Similarly, a disturbance in the surrounding air creates pressure waves in the air. These waves have a wave length and a frequency. The human ear can detect these pressure waves over a wide range of frequencies ranging from 20 cps to 10,000 cps (some people can detect frequencies up to 20,000 cps).

The term Hertz (Hz) is the unit that is used to indicate frequency in cycles per second. Thus, 1,000 Hz means 1,000 cycles per second frequency.

9-2 Sound Pressure

In the pond, the distance between the crests and troughs of the ripples represent the amplitude of the waves. In the air, the amplitude of a sound wave is represented by the change in pressure of the sound wave. Amazingly, the human ear can detect fluctuations in pressure that are as small as 1 billionth of a psi over a range where the larger pressures are a billion times larger than the smallest pressures.

Because of the extreme range of pressure levels that can be sensed by the human ear, a special "compressed" scale has been devised that reports sound pressure in terms of decibels (db). While it is not absolutely necessary to know exactly how much pressure corresponds to any given decibel level, it is important to know what decibel levels are acceptable to the human ear. Examples of decibel levels produced by several common sources is shown in the table below. Note that the 10 to 50 db levels define the range of acceptable noise levels for residential and light commercial applications.

Description Of Noise	Noise Level (db)
Threshold of pain	120
Rivet gun	95
Elevated train	90
Busy street traffic	70
Ordinary conversation	65
Quiet automobile	50
Quiet radio in home	40
Average whisper	20
Rattle of leaves	10
Threshold of hearing	0

9-3 Sound Power

Sound also involves energy. The voice of a person speaking to a group might generate between 0.000025 and 0.00005 watts; a tiny output to be sure. But sound power produced by equipment and machinery can vary over a considerable range and once again the decibel scale is used to express sound power levels.

Note that sound power level cannot be measured directly. However the sound power level can be related to the sound pressure level, which can be measured. Once the sound pressure level is measured, formulas can be used to predict the corresponding sound power level.

9-4 Frequency Spectrum

The sound generated by sources such as machinery or air turbulence is usually composed of several frequencies. In addition, the sound pressure (or sound power) level produced at each frequency may vary considerably. (An instrument called an "octave-band" analyzer is used to evaluate the db level in each frequency band.)

For example, a plot of decibel (db) level versus frequency for a fan (below) shows considerable variation of sound pressure (or power) with frequency. Note the db level for the fan is high (75 db) at lower frequencies (2000 Hz) and decreases at higher frequencies (20 db at 8000 Hz). Such a plot is called a frequency spectrum. (The frequency spectrum produced by a noisy air outlet would be the opposite of this fan spectrum, characterized by high db levels at high frequencies and low db levels at low frequencies.)

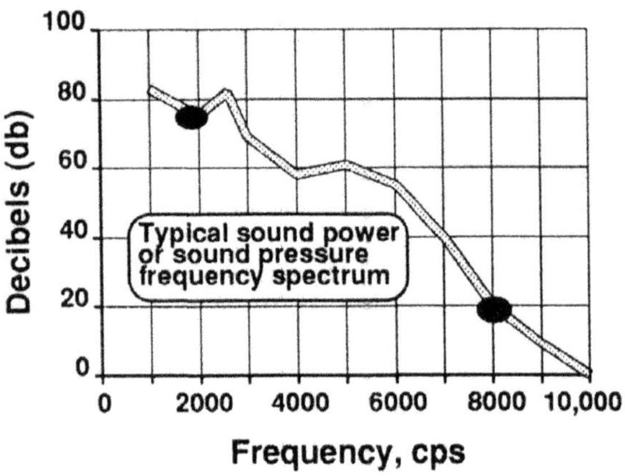

9-5 Loudness Chart

Frequency content has considerable influence on how loudness (of a sound) is interpreted by the human ear. At low db levels the ear is less sensitive to low frequencies than to high frequencies. As the db level increases, the ear tends to become uniformly sensitive to all frequencies. This is illustrated on the loudness chart below. Each curve represents the same sensation of loudness for the various frequency levels.

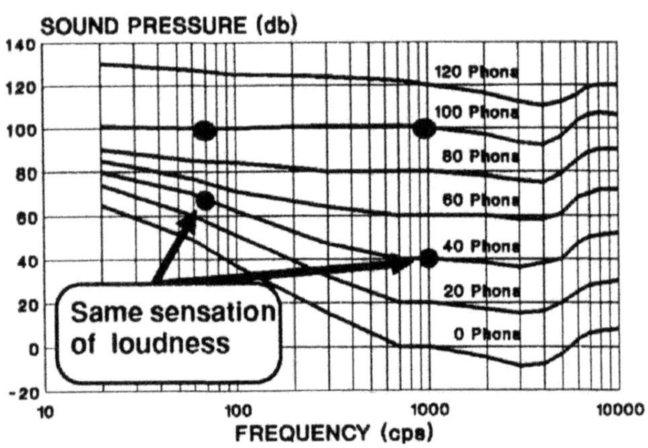

For example, a 50 Hz tone at 70 db is interpreted by the ear to be just as loud as a 1,000 Hz tone at (a mere) 40 db. But when the db level is increased to 100 db, the 50 Hz tone is interpreted as being just as loud as the 1,000 Hz tone. The point is that the db level must be evaluated and controlled at numerous frequencies to prevent unacceptable noise.

9-6 Noise Criteria Rating

Because the response of human ear is affected by both loudness and frequency, a number of register and grille manufacturers provide catalog performance data that includes the noise criteria (NC) ratings for the various models of outlets and inlets. This noise criteria rating corresponds to the way the human ear perceives loudness (refer to Section 9-5 above) because it defines specific pressure level limits (db) over the full range of hearing frequencies. An example of an NC chart is provided below.

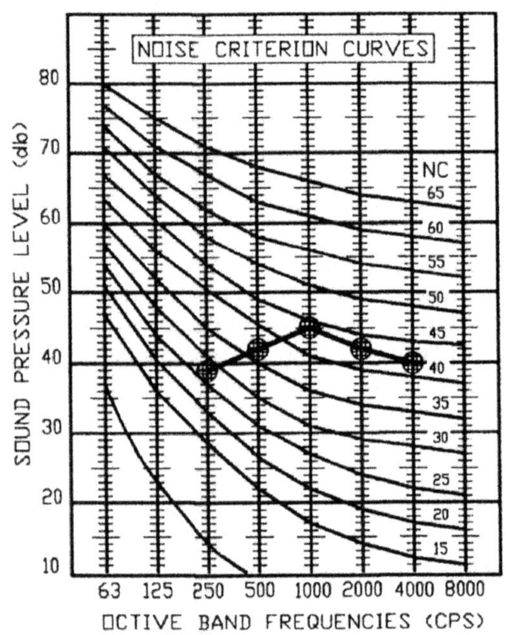

Through a series of tests, a register manufacturer can measure the db level for a particular register under a specific set of conditions such as air flow rate, velocity, approach, etc. These measurements can then be plotted on the NC chart and an NC rating for the register can be established. For example, the figure above shows five test points (db level and corresponding frequency) plotted on an NC chart. Notice that no db level (at any frequency) exceeds the NC-45 curve. Therefore the register used for this test receives an NC-45 rating and is so listed in the catalog.

9-7 Published NC Ratings

Many manufacturers publish performance data tables which tabulate the NC ratings that are associated with the various outlets (or inlets) that make up their product line. Note that for any particular device, the NC value increases as the rate of air flow (CFM) through the outlet (or inlet) increases. The table on top of the next page provides an example of air outlet performance data which tabulates the NC rating at various air flow rates.

Neck Velocity (fpm) Vel Press (In.Wg.)	400. 0.010	500. 0.016	600. 0.023	700. 0.062	800. 0.040	900. 0.051	1000. 0.063	1200. 0.090
Total Press (In.Wg.)	0.029	0.046	0.067	0.090	0.116	0.148	0.183	0.261
5 Inch Neck (CFM)	53	67	80	93	105	120	135	160
Throw (Ft)	2-2-5	2-3-6	2-3-7	3-4-8	3-5-8	3-5-9	4-6-9	5-7-10
NC	10	16	22	27	31	35	38	44
Total Press (In.Wg.)	0.038	0.061	0.087	0.114	0.152	0.194	0.24	0.342
6 Inch Neck (CFM)	80	100	120	140	160	180	200	235
Throw (Ft)	2-3-6	2-4-7	3-4-9	3-5-9	4-6-10	4-7-10	5-7-11	6-9-12
NC	12	18	25					
Total Press (In.Wg.)	0.049	0.078	0.113					
7 Inch Neck (CFM)	105	135	160					
Throw (Ft)	2-3-7	2-4-9	4-5-10					
NC	14	21	27					

Ceiling Diffuser Data

Note how the NC vaules increase as the CFM and neck velocity increase.

9-8 Acceptable NC for Various Applications

Some examples of of recommended NC levels for various applications are shown by the table below. Basically these recommendations take the ambient noise into consideration. Noise produced by the air conditioning equipment is acceptable as long as it is "masked" by the ambient noise.

Application	NC Level	Maximum Velocity
Private Residence		
Rural/Suburban	20 to 25	700
Urban	25 to 30	700
Apartments	30 to 35	700
Hotels/Motels		
Private room or suite	30 to 35	700
Meeting/Banquet room	30 to 35	700
Halls, Corredors, Lobbies	35 to 40	800
Service and support areas	40 to 45	900
Offices		
Executive	25 to 30	700
Conference rooms	25 to 30	700
Private	30 to 35	700
Open bay areas	35 to 40	800
Computer room	35 to 40	900
Business machine room	40 to 45	900
Waiting rooms	40 to 45	900
Retail Stores		
Clothing department	35 to 40	800
General sales areas	40 to 50	900
Supermarkets	40 to 50	900
Hospitals and Clinics		
Private/Semi-private rooms	25 to 30	700
Operating Rooms	25 to 30	700
Wards	30 to 35	700
Churches and Schools		
Sanctuaries	20 to 30	700
Libraries	30 to 35	700
Classrooms	30 to 35	700

9-9 Face Velocity Guidelines

Rather than providing specific NC ratings, many manufacturers provide face velocity or neck velocity guidelines for various applications. A listing of the maximum recommended velocity is also provided by the table below.

Application	NC Level	Maximum Velocity
Churches and Schools (cont)		
Laboratories	35 to 40	800
Gyms and rec. areas	40 to 50	900
Kitchens	40 to 50	900
Public Assembly		
Concert halls	20 to 25	700
Legitimate theaters	20 to 25	700
Auditoriums	25 to 30	700
Motion picture theatres	30 to 35	700
Libraries/Museums/Courts	30 to 35	700
Coliseums	30 to 35	700
Bank/Post office	35 to 40	800
Lobbies	35 to 40	800
Swimming pools	40 to 55	1000
Transportation Terminals		
Ticketing areas	30 to 35	700
General Waiting	35 to 45	800
Studios		
Recording studios	20 to 25	700
Broadcasting studios	20 to 25	700
TV audience studios	30 to 35	700
Restaurants/Cafeterias/Lounges		
Restaurant eating area	35 to 40	800
Lounges	35 to 45	800
Night clubs	35 to 45	800
Cafeterias	40 to 50	900
Industrial Production Areas	Background Noise Sets Limits	

Size	Area (SqIn)	CFM	40	60	80	100	125	150	175	200	225
		T.P. (In.Wg)	0.007	0.013	0.021	0.031	0.050	0.078	0.101		
8x6	34	Throw (Ft.)	3.5	5	6.5	8	11	15	19		
		Velocity	170	255	340	425.	530	635	745		
		T.P. (In.Wg)		0.007	0.012	0.02	0.028	0.041	0.056	0.083	0.109
10x6	43	Throw (Ft.)		4.5	6	7.5	10	12	15	18.5	21.5
		Velocity		200	265	335.	415	500	585	665	750
		T.P. (In.Wg)			0.007	0.01	0.021	0.029	0.042	0.060	0.081
12x6	53	Throw (Ft.)			4.5	6	8	10	12	14	16.5
		Velocity			220	275	340	410	480	560	615
		T.P. (In.Wg)				0.008	0.015	0.023	0.033	0.041	0.053
14x6	62	Throw (Ft.)				5	6	9	10.5	12	14
		Velocity				230	290	350	410	465	525

9-10 Catalog Velocity Recommendations

In some cases the catalog performance data may be published in such a way as to indicate a preferred selection range based on face or neck velocity. In the table above, selection should be made to the left of the heavy vertical bars to insure quiet operation.

While velocity is perhaps the most important criterion governing register noise, size is also a factor. Given two identical diffuser models, experiencing the same air velocity through the outlet, the larger outlet will generate more noise. This is also true for return air intakes. Thus, the general rule is to size larger returns based on a lower face velocity than smaller intakes.

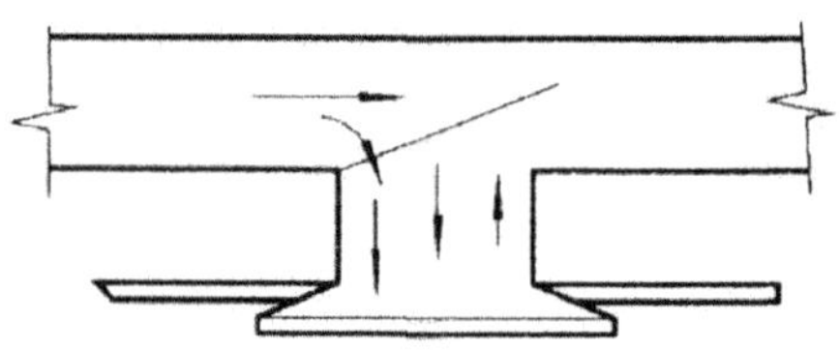

Case 1

Case 2

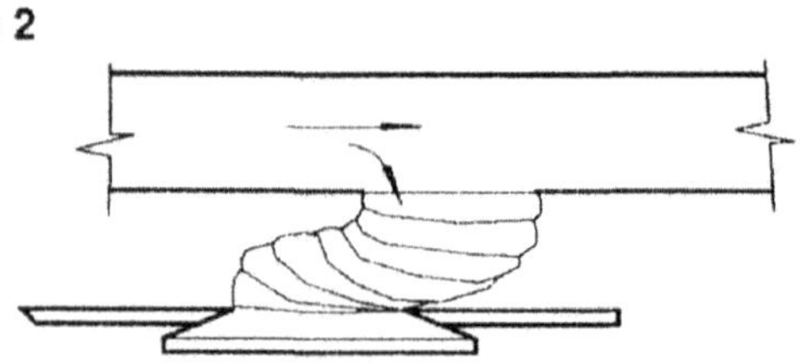

9-11 Installation Details Affect Noise Levels

The published noise level data represents sound measurements which were made in a laboratory with a uniform air velocity profile approaching the outlet. Field conditions can vary substantially from the test conditions. For example, non-uniform flow through the outlet (Case 1 below) will increase the generated noise. (Often, turning devices and equalizing grids are specified to reduce noise as well as to enhance performance.) Severely offsetting the outlet from the duct connection (case 2 below) can increase noise by as much as 10 to 20 db above the published rating.

9-12 Integral Dampers Generate Noise

Perhaps the most common and potentially greatest source of noise generated by a register is caused by the balancing damper immediately behind the grille. The amount of generated noise depends on how far the damper is closed. Closing this damper down severely can increase noise levels from 5 to 20 db above the published rating. (Note that register dampers are only intended for making minor adjustments to the flow rate. When appreciable flow reductions are required the flow adjustment should be made with hand dampers placed well upstream in the supply duct.)

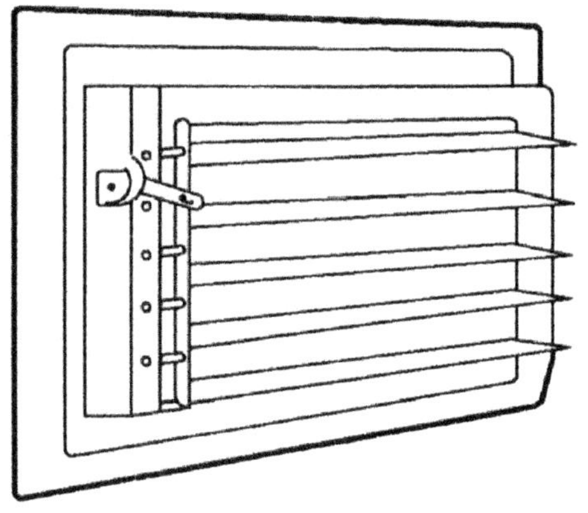

The chart below shows that air-side-generated noise is usually of a higher frequency than equipment-qenerated noise.

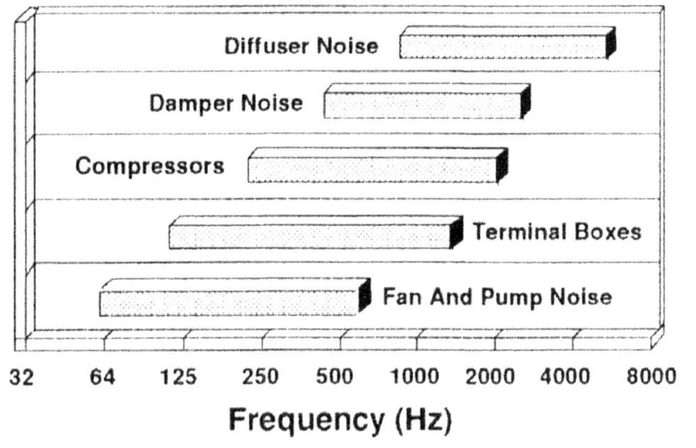

9-13 Attenuation

Elbows, take off fittings, transition, etc., help reduce (attenuate) the equipment generated noise that flows through the the duct system, eventualy entering the occupied space. For example, the noise generated by a remote (air handler) fan tends to flow through the duct system to the occupied space.

Duct fittings tend to attenuate equipment-generated noise before it reaches the occupied space. This is because fittings tend to reflect the sounds back upstream. Adding duct liner also helps especially if the insulation is placed directly in the elbow as well as a short distance up stream and down stream from the elbow.

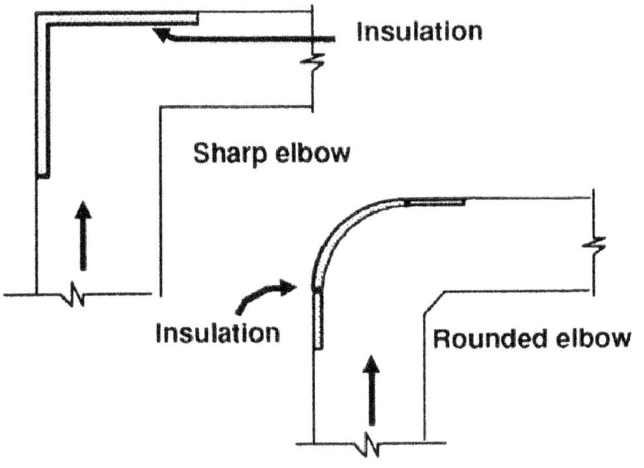

9-14 Generation

Lining a duct or an elbow cannot solve a noise problem if a register, fitting, damper or boot is the actual source of the noise. This strategy fails because, in this case the noise is generated downstream from the attenuating device, right at the terminal. Generally, if the noise is a rumble, it is the result of a fan or blower. A whistle on the other hand is a result of high frequency sound. If the noise disappears when the outlet or inlet is removed, obviously the terminal is to blame.

9-15 Equipment Near the Occupied Space

Furnace closet applications are probably the most critical duct system application in homes, offices, apartments, etc., in terms of noise abatement. Usually, the closet is adjacent to the occupants and the slightest noise can be disturbing or at least distracting.

This type of installation (below) often features a very short, single return connection. This short return connection can be a major cause of noise transmission. Even though the air is moving inward, the equipment- generated sound can radiate outward with the short duct acting like a cheerleader's megaphone. The intake grille as a cover is of modest help in reducing equipment noise transmission.

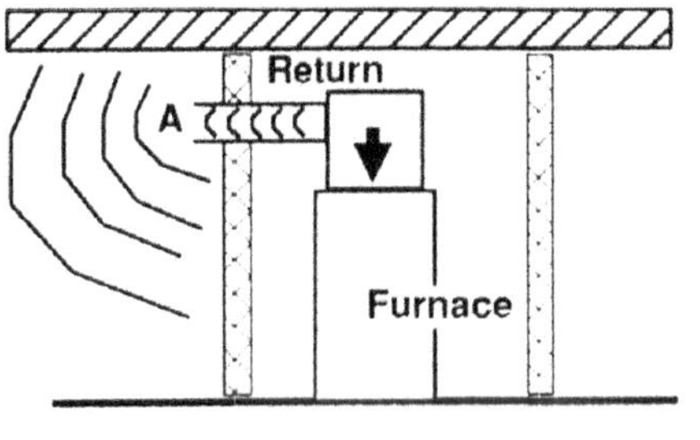

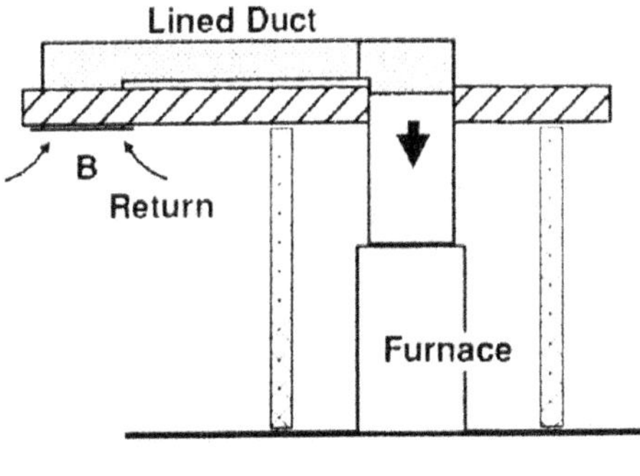

The best and recommended solution, is to avoid short returns completely. As illustrated above, the return intake should be sized for a very low face velocity, the return duct should be lengthened and lined, plus the air should be made to turn at least twice in its path toward the equipment. The fan should, of course, be adequately sized and rotating at moderate speeds to minimize the sound power level at its origin.

Note that sound can also enter the occupied space by transmission through equipment room walls. In this case make sure that the wall between the equipment room and the occupied space is completely sealed (no crackage or penitrations). Sound absorbing material, or in extreme cases, additional wall mass, may be rerquired If after the wall is sealed, the transmitted noise is still a problem.

Section 10

Selecting Supply Outlets

The supply outlet selection and sizing process is critical. Improper air terminal selection can result in inadequate heating and/or cooling performance in some or all rooms. Poor temperature control, drafts, stagnant air and noise are some examples of the problems that can result when the wrong types or sizes of supply outlets are used.

10-1 Objectives

The following list summarizes the various objectives that should be kept in mind when the supply outlets are selected.

- No stagnate air in the occupied zone.
- Velocities in occupied zone between 20 fpm and 50 fpm.
- Supply air "mixes and circulates" with the room air.
- Acceptable noise level.
- Offset drafts moving down cold windows or walls.
- No collisions between high velocity air streams.
- No collisions between high velocity air and walls or ceiling.
- No "short circuit" between supply and return terminals.

10-2 Location

Supply air terminal location must also be compatible with the architectural design, thermal characteristics of the envelope and the overall HVAC system design.

- Ceiling and high (interior) sidewall terminals can be used when duct runs are located in the attic (or on roof).
- Floor and low sidewall terminals can be used when duct runs are located in basement or underneath a ground slab.
- Perimeter floor or base board terminals are preferred in cold climates specifically for slab construction. (This is less important if the structure is extremely well insulated and has a basement.)
- Low vertical discharge terminals (wide spread) installed under windows and/or glass doors tend to offset cold drafts and convection currents.
- High terminals are better for cooling (hot climates) and low terminals are better for heating (cold climates).

10-3 Approximate Air Terminal CFM

The CFM delivered by a single terminal will depend on the load and the number of terminals that are required to properly introduce the supply CFM into the room. In commercial work a single terminal may deliver less than 100 CFM or more than a 1000 CFM. In residential work a single terminal is usually limited to less than 200 CFM. The approximate supply CFM that is required to cool or heat a room can be estimated by using the "sensible heat" equation (see below).

$$CFM = \frac{\text{Room Sensible Cooling Load (Btuh)}}{1.1 \times (\text{Room db} - \text{Supply Air db})}$$

$$CFM = \frac{\text{Room Heating Load (Btuh)}}{1.1 \times (\text{Supply Air db} - \text{Room db})}$$

10-4 Selection Procedure

The location and style (ceiling, floor, side wall etc.) are usually dictated by the architecture of the structure and the physical arrangement of the HVAC equipment (refer to the paragraphs above). Model and size are determined by referring to manufacturer's performance data which provides information on CFM, throw, spread and noise.

Step 1 - Scan the performance data for a size that is capable of delivering the desired CFM. There usually will be more than one size that meets these requirements.

Step 2 - Make sure that the noise level is acceptable. If NC values are published, select a model and size that has an acceptable NC rating. If NC values are not published, base the selection on face velocity or neck velocity. Choose models and sizes that have a face velocity or neck velocity that is less than 700 fpm. Refer to the tables in Section 10-6 for a listing of recommended face velocities and NC values for various applications. (The 700 fpm value can be exceeded in applications where terminal generated noise is not a consideration.)

Step 3 - Check the throw. Note that the throw is affected by the spread, less spread usually results in a longer throw. Also note that the catalogued throw corresponds to some specific terminal velocity. Study the catalogue data to determine the magnitude of the terminal velocity. Throws terminating in the occupied zone should not exceed 50 fpm. Terminal velocities at throw boundaries such as walls, ceilings or adjacent air patterns can range between 50 fpm and 150 fpm.

Step 4 - Check the pressure drop. (Pressure drop information is required for residential duct sizing calculations or for commercial fan sizing calculations.)

10-5 Effect of Fan Operation

Supply outlets must perform adequately at all points of blower operation. Variable volume systems (two speed, multiple speed, variable speed, inlet vanes etc.) usually require a special type of outlet which is designed to provide an acceptable air pattern over a wide range of flow rates.

- Constant volume terminals will only perform satisfactory over a relatively narrow range of air flow rates. When fan CFM varies. say for summer and winter operation, it is necessary to check the performance (throw, spread, velocity and noise) at the maximum and minimum CFM values. Two constant volume supply outlets will be required when the operating range of a single outlet is inadequate. (Dampered runouts which feed the second set of outlets can be used for change over.)

- Variable volume terminals are preferred when the supply CFM modulates. These terminals are usually located at the ceiling and they provide a remarkably wide operating range by taking advantage of the surface effect.

10-6 Noise Criteria

High face or neck velocities can result in objectionable noise. The outlet selection process must insure that outlet generated noise will be within acceptable levels. The table below provides a listing of acceptable NC values and a listing of recommended face or neck velocities.

Application	NC Level	Maximum Velocity
Private Residence		
Rural/Suburban	20 to 25	700
Urban	25 to 30	700
Apartments	30 to 35	700
Hotels/Motels		
Private room or suite	30 to 35	700
Meeting/Banquet room	30 to 35	700
Halls, Corredors, Lobbies	35 to 40	800
Service and support areas	40 to 45	900
Offices		
Executive	25 to 30	700
Conference rooms	25 to 30	700
Private	30 to 35	700
Open bay areas	35 to 40	800
Computer room	35 to 40	900
Business machine room	40 to 45	900
Waiting rooms	40 to 45	900
Retail Stores		
Clothing department	35 to 40	800
General sales areas	40 to 50	900
Supermarkets	40 to 50	900
Hospitals and Clinics		
Private/Semi-private rooms	25 to 30	700
Operating Rooms	25 to 30	700
Wards	30 to 35	700
Churches and Schools		
Sanctuaries	20 to 30	700
Libraries	30 to 35	700
Classrooms	30 to 35	700

10-7 Manufacturer's Data

The Air Diffusion Council publishes a test code (ADC 1062: GRD-84) for grilles, registers and diffusers. Some of the more important requirements of this code are listed below. (Isothermal tests are conducted with a zero-degree temperature difference between the supply air and the room air.)

Standard ADC test code reporting requirements:

- Static pressure drop vs CFM
- Total pressure drop vs CFM
- Sound (NC) vs CFM
- Isothermal throw (150 fpm terminal velocity) vs CFM
- Area factor (Ak)

Optional ADC test code reporting requirements:

- Nonisothermal throw tests, horizontal projection
- Nonisothermal throw tests, vertical projection
- Nonisothermal throw tests, projection at an angle
- Drop tests, horizontal projection
- Drop tests, angled projection

It is important to realize that the requirements of the ADC test standard are somewhat flexible. This flexibility results in noticeable differences in the way that manufacturers report on outlet performance. For example, any given manufacturer may or may not report on ceiling diffuser or sidewall outlet drop; or different manufacturer's may list throws at different terminal velocities; or any given manufacturer may choose,

Application	NC Level	Maximum Velocity
Churches and Schools (cont)		
Laboratories	35 to 40	800
Gyms and rec. areas	40 to 50	900
Kitchens	40 to 50	900
Public Assembly		
Concert halls	20 to 25	700
Legitimate theaters	20 to 25	700
Auditoriums	25 to 30	700
Motion picture theatres	30 to 35	700
Libraries/Museums/Courts	30 to 35	700
Coliseums	30 to 35	700
Bank/Post office	35 to 40	800
Lobbies	35 to 40	800
Swimming pools	40 to 55	1000
Transportation Terminals		
Ticketing areas	30 to 35	700
General Waiting	35 to 45	800
Studios		
Recording studios	20 to 25	700
Broadcasting studios	20 to 25	700
TV audience studios	30 to 35	700
Restaurants/Cafeterias/Lounges		
Restuarant eating area	35 to 40	800
Lounges	35 to 45	800
Night clubs	35 to 45	800
Cafeterias	40 to 50	900
Industrial Production Areas	Background Noise Sets Limits	

as an option, to report on nonisothermal performance. The net result is that there are as many different ways to publish outlet performance data as there are manufacturers. With this in mind it becomes absolutely necessary to study manufacturer's performance data, (and any footnotes) carefully to determine exactly what the test data really means. A summary of the ADC test procedure requirements is provided below.

ADC Test Procedure Requirements:

- The test room shall have a 9-foot high ceiling.
- All cone and vane adjustments shall be noted on the data sheet.
- Ceiling diffusers shall be installed flush on the ceiling surface.
- Wall grilles and registers shall be installed so that the top of the outlet is 9 inches below the ceiling.
- Throw tests shall be conducted with isothermal air.
- Nonisothermal throw tests shall be conducted with a room air to supply air temperature difference of 20 degrees.
- Throw tests can be conducted at any appropriate terminal velocity provided that the test velocity is noted on the data sheet.
- Drop tests shall be conducted with a room air to supply air temperature difference of 20 degrees.
- The throw that is associated with a ceiling diffuser that is not mounted near the ceiling can be approximated by multiplying the flush mounted throw by 0.707 adjustment factor.
- The throw that is associated with a high sidewall outlet that is not mounted near the ceiling can be approximated by multiplying the "near ceiling" throw a "throw reduction factor" that has been verified by a laboratory test.
- Pressure tests (outlet only or outlet-accessory combination) shall establish the minimum static pressure at the tested flow rate.

10-8 Floor, Baseboard and Low Sidewall Outlets

Floor, baseboard and low sidewall outlets are excellent for heating applications and they are reasonably effective for cooling applications. This class of outlets are ideal for heating and cooling buildings located in cold climates and they are particularly effective for heating ground slab buildings.

Throw:
During heating an isothermal throw of from 4 to 6 feet is required when the outlet delivers the design air flow rate (CFM). Cooling requires a longer throw to project the cooler and denser air higher into the room. A isothermal throw of from 6 to 8 feet is recommended for cooling, again at the design air flow rate.

Spread:
Wide blanketing spreads are desirable, but in many cases the spread is limited by the throw requirements.

Face velocity:
High face velocities can result in objectionable noise. In general, face velocities should be less than 600 fpm with 700 fpm as an upper limit. Refer to section 10-6 for a more exact definition of acceptable face velocities.

Supply to room temperature difference:
During heating supply air to room air temperature differentials can range up to 75 F without causing problems. During cooling room to supply air temperature differentials of 20 F or less can be accommodated without resorting to selections that result in unreasonably high face velocities.

Pressure drop:
The static pressure drop should be noted because it will be required for duct sizing and/or fan selection. Make sure that the pressure drop data corresponds to the selected hardware combination. (The pressure drop across a register is greater than the pressure drop across a grille because of the additional pressure drop across the register's damper.) Also note that an additional pressure drop may be associated with any devices, boots or fittings that are packaged with the outlet. When only the total pressure drop is published, the static pressure drop can be estimated by subtracting the up stream velocity pressure from the published total pressure value.

10-9 Example: Floor or Baseboard Outlet Selection

Refer to the examples of manufacturer's floor and baseboard diffuser performance data provided in appendix A1. Select various types of floor and baseboard diffusers for the room and the conditions described by the figure below.

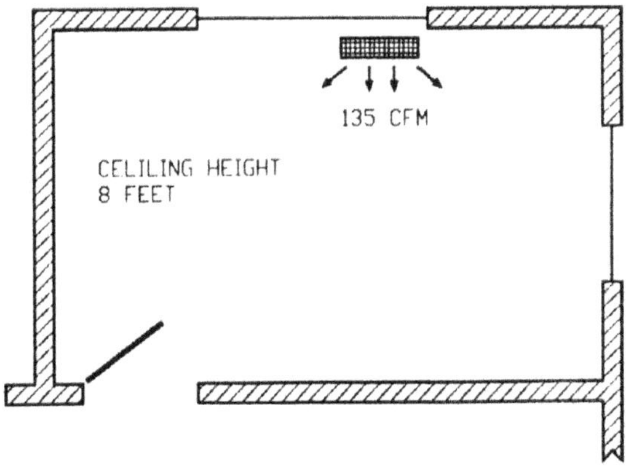

The table below summarizes the performance characteristics of a few of the outlets which qualify. Just considering performance, the models with longer throws, wider spreads and lower face velocities are preferred.

Model	Type	Size	Throw	T.vel	F.vel	Spead	Static
411	floor	4x12	6	50	600	10.5	0.022
421	floor	4x12	8	50	600	9.5	0.022
531	floor	6x14	9	50	400	5.0	0.010
406	basbr	2x14	8	50	475	7.5	0.032
464	basrd	4 ft	8	50	347	13.5	0.043

Since (in this example) the pressure drop data does not specify whether the data corresponds to the total pressure drop or the static pressure drop assume (conservatively) that the data corresponds to the static pressure loss. Since the picture and description indicates a grille/damper combination the pressure drop data probably includes the pressure drop for the damper.

10-10 Ceiling Diffusers

Ceiling mounted diffusers are excellent for cooling applications but they are less effective for heating applications. Ceiling diffusers are ideal for cooling and heating buildings located in mild climates or for supplying air to interior rooms or zones. Ceiling diffusers are a poor choice for heating ground slab buildings that are located in cold climates.

Throw:

In general, ceiling diffuser throw data corresponds to the situation where the outlet is mounted flush on a 9- foot high ceiling. If the outlet is mounted away from the ceiling the surface effect will not take affect and the throws can be reduced by as much as 30 percent (0.707 correction factor). Study the published data, verify the mounting details and/or search for correction factor information.

Published values for throws or diffusion radii usually correspond to a isothermal test condition (supply air temperature is the same as the room air temperature). Corresponding terminal velocities may range from 50 fpm and 150 fpm. Isothermal terminal velocities in this range usually insure that the velocities in the occupied zone will be fall between 20 fpm and 50 fpm.

When isothermal performance is the only data provided, a rule of thumb is that the isothermal throw at 100 fpm is approximately the same as the 50 fpm throw with a 20 degree (cooling) temperature difference. Therefore, when isothermal data is all that is available, the outlet selection should be based on the isothermal throw that corresponds to a 100 fpm terminal velocity.

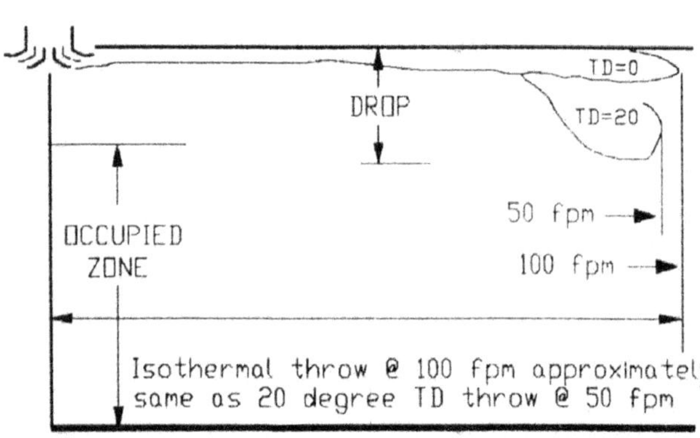

The throw (or radius of diffusion) for ceiling diffusers with a 360 degree pattern should be equal to the distance between the diffuser and the nearest wall or, when there are two or more ceiling outlets serving the space, equal to 1/2 the distance between adjacent outlets. (Based on 100 fpm terminal velocity.)

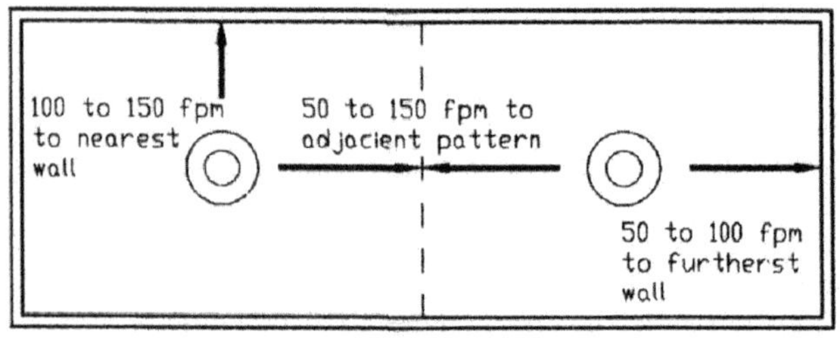

For directional ceiling diffusers acceptable throw ratios range from 0.75 to 1.2 times the distance to the nearest wall to which the outlet is projecting. During heating a throw ratio of 1.0 to 1.2 is preferred because the size of stagnate air zone will be reduced. (Based on 100 fpm terminal velocity.)

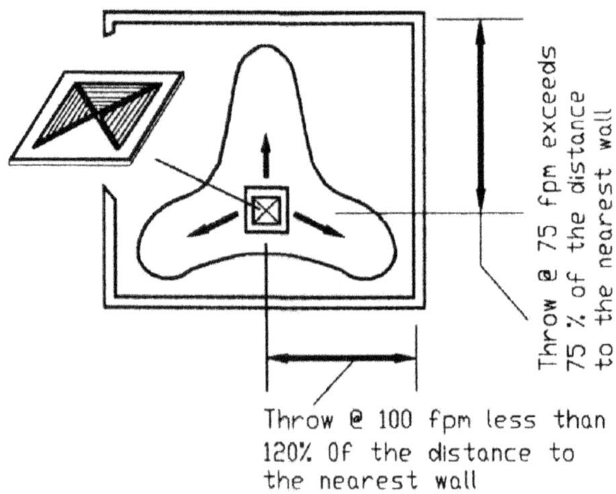

Neck velocity:

High neck velocities can result in objectionable noise. For residential and many commercial applications, neck velocities should be 700 fpm or less. In applications where noise is not a primary consideration, a 900 fpm neck velocity can be used as an upper limit. The surest way to insure that the neck velocity will not result in objectionable noise is to verify that the published NC number (or neck velocity) is less than the NC value (or neck velocity value) recommended by the table that appears in Section 10-6.

Drop:
Manufacturers may or may not publish "drop" information. Normally drop will not be a problem when the outlet is mounted flush to the ceiling and the manufacturer's throw recommendations are followed. However, drop can be a problem if ceiling terminals are mounted away from the ceiling. In this case high velocity unmixed air may be projected into the occupied zone.

Supply to room temperature difference:
Supply air to room air temperature differentials affect throw and drop. (Refer to section 4-10.) During cooling, differentials that fall between 10 and 25 degrees can usually be handled without causing problems. Actual performance depends on the ceiling mounting height and the temperature differential. Unfortunately comprehensive data which cross references ceiling height, temperature difference, throw and drop is not usually published. In general a greater number of smaller terminals should be used with low ceilings and a fewer number of larger terminals should be used with higher ceilings. The following table can be used as a guide to estimate the maximum CFM that should be delivered by a single outlet.

Maximum CFM Per Ceiling Outlet

Temp Diff	Ceiling Height (Ft)						
(F)	8	9	10	11	12	14	16
10	600	1500	4000				
15	425	1150	2250	4400			
20	250	700	1400	2800	4500		
25	175	500	975	1800	3000	6000	
30	100	200	550	1000	1500	3500	6500

During heating, the buoyancy affect can be minimized by using the smallest possible supply air to room air temperature differential. In this case a 25 F differential is acceptable for directional diffusers and 20 F differential is preferred for circular diffusers. This means that typically, heating supply air should be about 100 F.

Pressure drop:
The static pressure drop should be noted because it will be required for duct sizing and/or fan selection. Make sure that the pressure drop data corresponds to the selected hardware combination. (The pressure drop should include the pressure drop that is associated with turning and straightening devices such as extractors, equalizing grids, etc..) When only the total pressure drop is published, the static pressure drop can be estimated by subtracting the up stream velocity pressure from the published total pressure value.

10-11 Example: Nondirectional Ceiling Diffuser

Refer to the examples of manufacturer's non-directional ceiling diffuser performance data provided in Appendix A1. Select non-directional (circular pattern) ceiling diffusers for the room and the conditions described by the figure below.

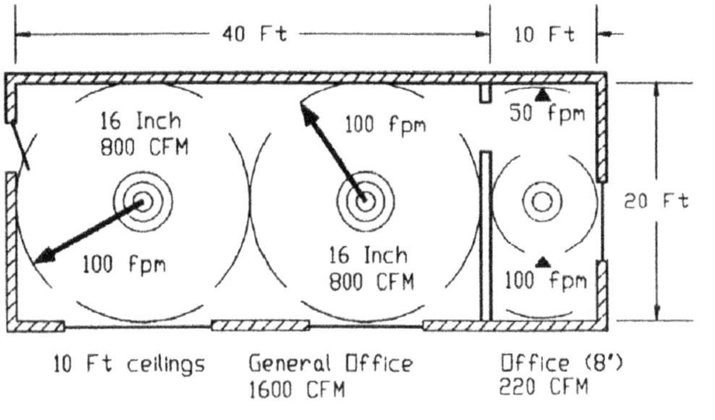

In general it is desirable to use as few diffusers as possible while still maintaining performance within an acceptable range of throw boundaries and noise levels. In this example adequate performance is accomplished with two diffusers (see previous figure) in the general office and with one in the private office.

The table below summarizes the performance characteristics of round diffusers and square diffusers. In both cases the terminal velocities at the throw boundaries fall between 50 fpm and 100 fpm and the NC levels are well below the recommended levels. The static pressure was calculated by subtracting the velocity pressure value from the total pressure value. (Since these particular data sheets do not mention dampers or equalizing grids assume that the static pressure drop corresponds to a terminal with no accessories.)

General Office:

Model	Type	Size	Throw	T.vel	N.vel	NC	Static
TMR	Rn'd	16"	9.5	100	550	19	0.045
TMS	Sq'r	15"	8.5	100	650	24	0.021

Private Office:

Model	Type	Size	Throw	T.vel	N.vel	NC	Static
TMR	Rn'd	8"	5.0	100	600	21	0.052
TMS	Sq'r	8"	5-0	100	650	15	0.035

10-12 Example: Directional Ceiling Diffuser

Refer to examples of manufacturer's directional ceiling diffuser performance data provided in Appendix A1. Select a directional ceiling diffuser for the room and the conditions described by the figure below.

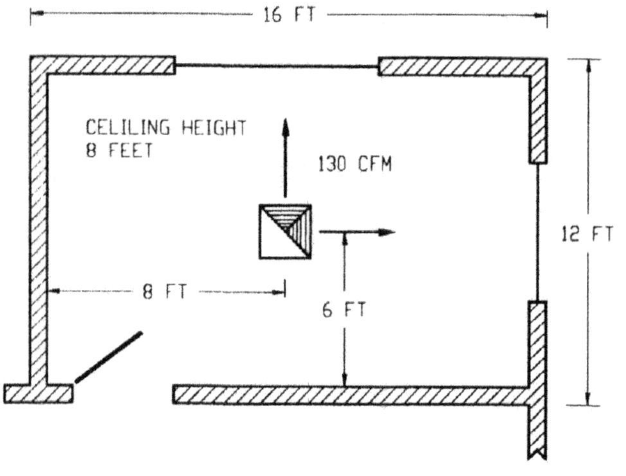

In general it is desirable to take advantage of the directional discharge design by selecting a diffuser which has an air pattern that will throw the supply air toward the outside walls. The table below summarizes the performance characteristics of a few of the outlets which qualify. However, the A300 series is preferred because it's (two-way and three-way) throws can be oriented at 90 degrees to each other, which is ideal for this room. (Since the C-series outlet only throws left and right it is not able to blanket both outside walls.)

Model	Type	Size	Throw	T.vel	F.vel	NC	Static
A-302	2way	12x12	8.0	75	350	na	0.008
A-303	3way	10x10	8.0	75	500	na	0.016
C-type	2way	14x6	6.0	75	600	na	0.022

Note that the catalog pressure drop data does not specify whether the data corresponds to the total pressure drop or the static pressure drop. Therefore assume (conservatively) that the data corresponds to a static pressure loss. Since the catalog description indicates a grille/damper combination the pressure drop data probably includes the pressure drop for the damper.

10-13 High Sidewall Outlets

High sidewall grilles and diffusers are well suited for cooling applications and, providing they have adequate throw, can be used successfully for heating applications. However, during cooling, high sidewall outlet performance must be carefully evaluated to insure that drafts are not caused by excessive drop.

Throw:

Certified (as per the ADC test code) high side wall outlet throw data corresponds to the situation where the top of the outlet is mounted near the ceiling. Study the manufacturer's data carefully to verify what mounting arrangement was used to generate the performance data. When an outlet is mounted away from the ceiling the throw will be reduced. In this case the throw can be estimated if manufacturer provides the appropriate adjustment factor.

Also note that published throw data usually correspond to the isothermal condition and that the corresponding terminal velocities may range from 50 fpm to 150 fpm. In some cases the manufacturer may choose to supply nonisothermal throw data. In any case, it is best to use the data that provides the closest match to the conditions that will be expected to occur at the site.

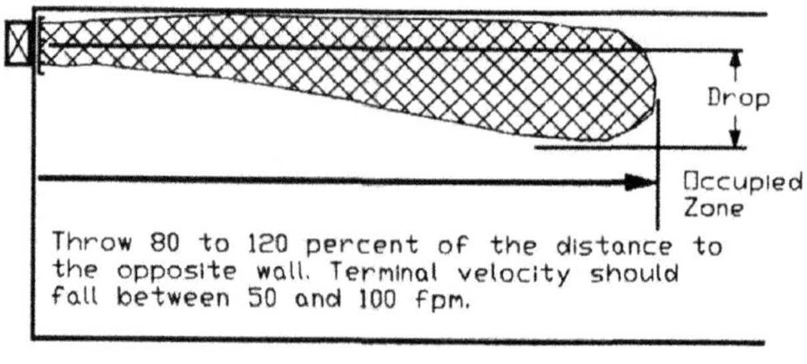

For cooling, acceptable isothermal throw ratios can vary from 0.75 to 1.2 times the distance to nearest wall. However throws ratios of 1.0 to 1.2 are preferred because they reduce drop during cooling and tend to reduce the size of the stagnate zone during heating. (Throw recommendations correspond to a terminal velocity of 100 fpm.) Note that throws can be increased by adjusting the horizontal vanes so that they deflect the primary air stream up toward the ceiling.

Drop:

It is important to determine the approximate drop because excessive drop could cause drafts in the occupied zone. Unfortunately many manufacturers do not publish drop data. If drop data is available, use it. If drop data is not available refer to the (generic) drop graphs provided below.

The first drop chart provides information on the drop that can be expected to occur when the high sidewall register is mounted close to the ceiling with the vanes adjusted for horizontal discharge and a narrow (zero to twenty degree vane angle) spread.

High Sidewall Near Ceiling
Horizontal Throw
0 To 20 Degree Spread

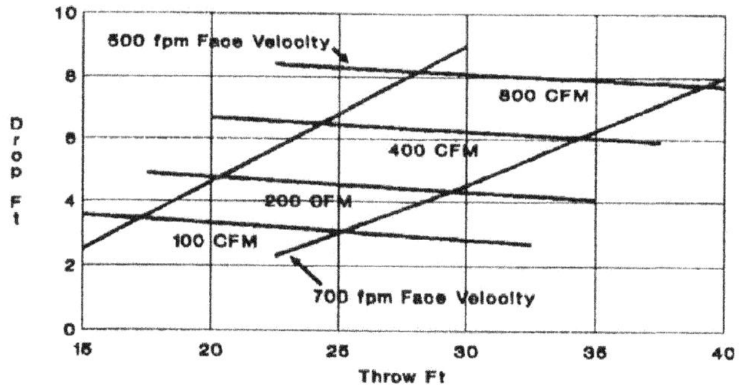

The next drop chart provides information on the drop that can be expected to occur when the high sidewall register is mounted close to the ceiling with the vanes adjusted for horizontal discharge and a wide (30 to 45 degree vane angle) spread.

High Sidewall Near Ceiling
Horizontal Throw
30 To 45 Degree Spread

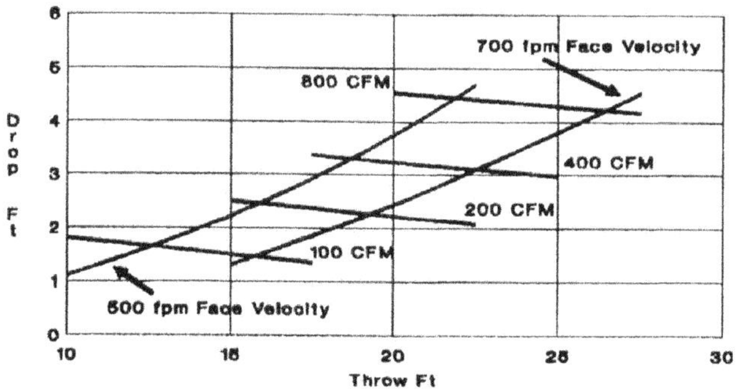

The final drop chart provides information on the drop that can be expected to occur when the high sidewall register is mounted 2 to 4 feet below the ceiling with the vanes adjusted to discharge the air at a 20 degree upwards angle and a narrow (zero to twenty degree vane angle) spread.

High Sidewall Below Ceiling
20 Degrees Up
Zero Degree Spread

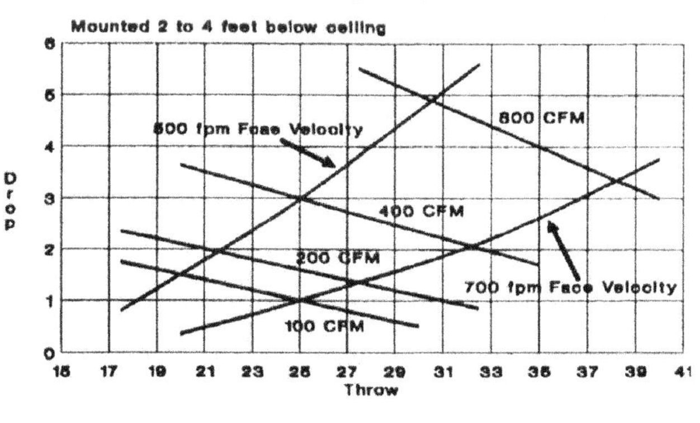

Refer to the three charts above and note that when the outlet is mounted near the ceiling the drop can be significantly reduced by increasing the spread. In general it is better to use several small outlets with a wide spread instead of a few larger outlets with a narrow spread. When drop is excessive, the drop can be reduced by installing the outlet 2 to 4 feet below the ceiling with the horizontal vanes set (20 degrees) to deflect the air up toward the ceiling.

Spread:
In addition to throw and drop, the spread of the discharging air pattern must also be considered when selecting high sidewall outlets. The spread should not exceed the available space between outlet center-lines or the adjacent walls. If the spread is too wide adjacent air streams will merge. Sidewall outlets should be spaced so that the air streams do not merge until the air has been projected to at least 50 percent of the throw (see the following figure). When air streams merge beyond the 50 percent point there will be a slight increase in throw, perhaps 15 percent or so.

Most register manufacturers provide spread versus throw data for selected vane settings. This data can be very useful when selecting and applying high sidewall registers. When they are drawn to scale on a floor plan, the spread patterns can be used to verify outlet performance.

Air patterns that are sketched on scale drawings of the room allow the designer to determine if the spreads, throws and merge points are acceptable. The air patterns can be approximated by a set of diverging lines. The following diagram at the top of the right hand column (above) shows that the included angle between the lines which are tangent to the air pattern is equal to either 20 degrees, 30 degrees or 60 degrees depending on the vane setting.

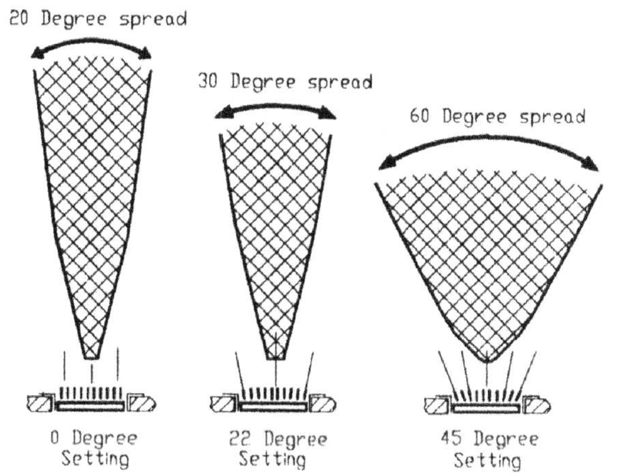

Face velocity:
High face velocities can result in objectionable noise. In general, face velocities should be less than 600 fpm with 700 fpm as an upper limit. Refer to Section 10-6 for a more exact definition of acceptable face velocities.

Supply to room temperature difference:
Supply air to room air temperature differentials affect throw and drop. (Refer to section 4-10.) During cooling, differen-

tials of 20 degrees or less can usually be handled without causing problems. Actual performance depends on the ceiling height, mounting height, temperature differential and vane settings. Unfortunately comprehensive data which cross references ceiling height, mounting height, temperature difference, vane setting, throw and drop is not usually published. In general a greater number of smaller terminals should be used with low ceilings and a fewer number of larger terminals should be used with higher ceilings.

During heating, the buoyancy effect can be minimized by using the smallest possible supply air to room air temperature differential. A 20 to 25 F differential is desirable but larger differentials are acceptable.

Pressure drop:
The static pressure drop should be noted because it will be required for duct sizing and/or fan selection. Make sure that the pressure drop data corresponds to the selected hardware combination. (The pressure drop across a register is greater than the pressure drop across a grille because of the additional pressure drop that is associated with the register's damper.) Also note that an additional pressure drop may be associated with any devices, boots or fittings that are packaged with the outlet. When only the total pressure drop is published, the static pressure drop can be estimated by subtracting the up stream velocity pressure from the published total pressure value.

10-14 Example: High Sidewall Outlet Selection

In general it is desirable to use as few outlets as possible while still meeting the throw, drop and noise level requirements. In addition, the location of the supply trunk (duct) also has a direct effect on the size and location of the outlets. In the examples below three duct locations and supply patterns are analyzed.

Case One:
Trunk duct runs along the long side of the room

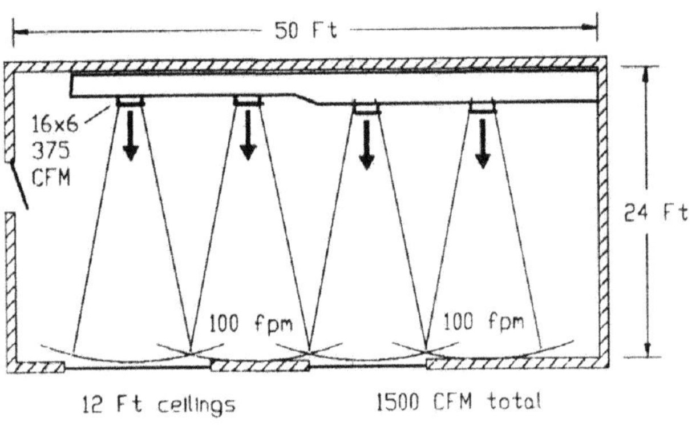

Size	Spr'd	Drop	Throw	T.vel	F.vel	NC	Static
16 x 6	22	5	20.0	100	700	24	0.068

Case Two:
Trunk duct runs in a soffit down the middle of the room

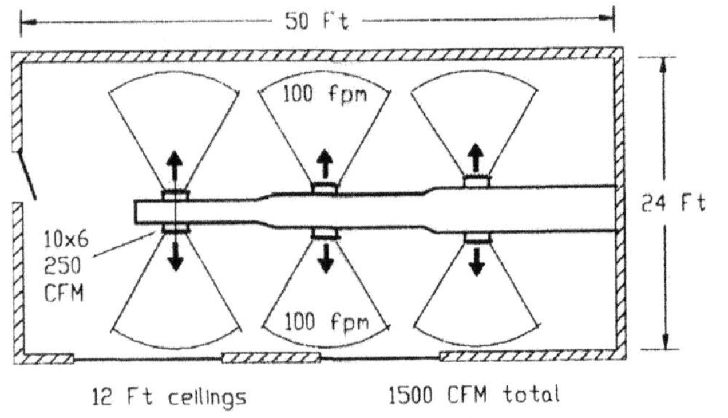

Size	Spr'd	Drop	Throw	T.vel	F.vel	NC	Static
10 x 6	45	3	10.0	100	700	21	0.119

Case Three:
Trunk duct runs along the short side of the room

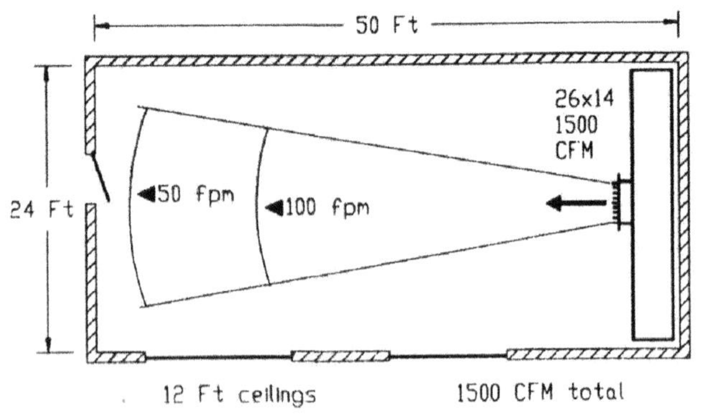

Size	Spr'd	Drop	Throw	T.vel	F.vel	NC	Static
26 x 14	22	6	38.0	100	600	25	0.051

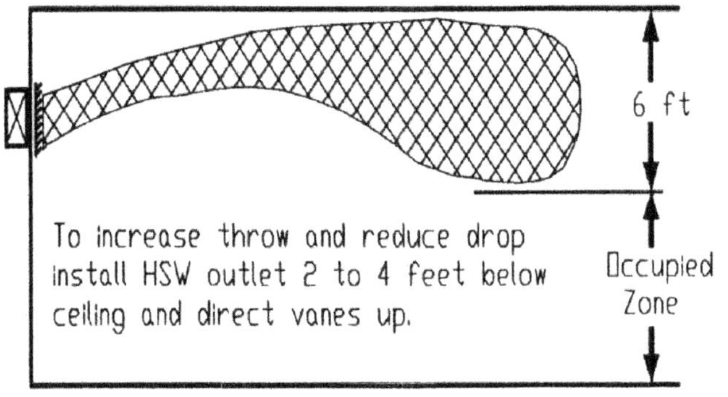

The figures above summarize the air patterns and performance characteristics that are associated with the various arrangements and outlet selections. The best overall performance was obtained by using six outlets located along the center of the room, but in all cases the throws, merge points, drops and NC levels are acceptable. Drops were estimated by using the generic drop charts provided in the section above. Static pressure was calculated by subtracting the velocity pressure value from the total pressure value. Since the catalog (picture and description) indicates a grille/damper combination, assume that the pressure drop data includes an allowance for the damper.

Section 11

Selecting Return Inlets

There are as many types of return inlet designs as there are supply outlet designs (ceiling, floor, sidewall, round, square, slot, perforated, etc. etc.). However, selecting a return inlet is consideably simpler than selecting a supply outlet. Just remember that the return system must provide an unobstructed path from each and every room or zone back to the equipment. The number of returns that are required for any system will depend on how many rooms or zones are classified as separate, isolated areas. Return system performance will always be acceptable as long as there are enough returns of adaquate size.

11-1 Style and Location

The style is dictated by the architecture and by the location of the return duct system. Square, round, perforated or slot designs are usually used in conjunction with return ducts which are installed above the ceiling. Sidewall designs are usually associated with risers or drops. Floor, baseboard and low sidewall designs are usually used in conjunction with return ducts which are installed below the floor.

Location in a stagnate zone is desirable but not critical. Never locate a return in a position which establishes a path that allows the supply air to short circuit directly into a return.

11-2 Size

Return air grilles should be sized to keep inlet-generated noise to a minimum. In residential work the return size is usually based on achieving either a 300 fpm velocity over the gross area of the grille (face velocity) or a 400 fpm velocity over the free area of the grille (core velocity). In commercial applications the return size may be based on 400 fpm face velocity or 500 fpm core velocity or it can be based on a noise criterion (NC) rating. Refer to section 9-8 or 10-6 for guidlines on acceptable NC values.

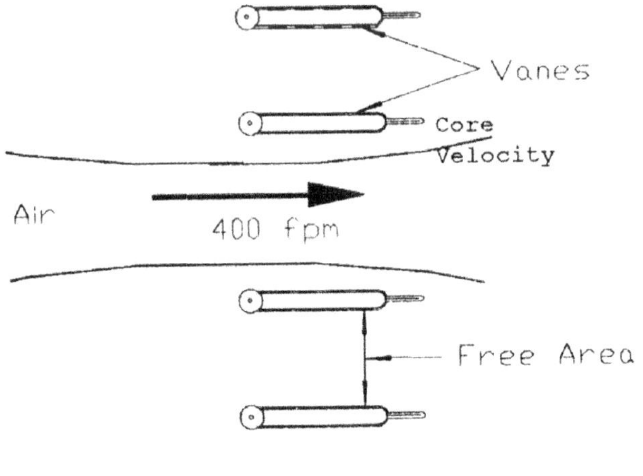

11-3 Pressure Drop

In all cases, the pressure loss of the intake should be noted and recorded because this information will be required for the duct design. In residential systems it is unusual for the pressure drop to exceed .03 inches w.g. when the previously noted velocity limits are observed. In commercial work it is possible to select a size that meets the NC requirement and then find that the pressure drop through the return is larger than desired. In these cases it is advisable to select a larger return in order to reduce the pressure drop.

11-4 Transfer Grilles

Transfer grilles can be used to combine isolated rooms or zones into a single larger return zone which can be served by a single return inlet. Because transfer grilles are not directly connected to the return duct they should be sized at a lower velocity.

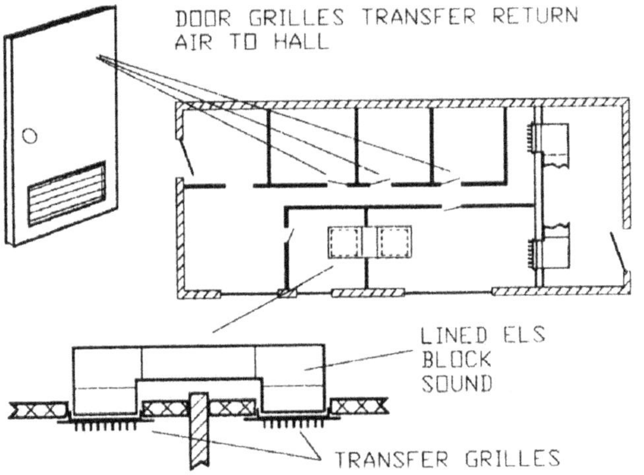

Transfer grilles are typically sized at 200 fpm face velocity or 300 fpm core velocity and in some cases a value of 150 fpm face velocity is used to minimize the pressure drop. If door undercuts are used in lieu of door grilles, allow 1 inch of undercut for each 60 CFM of return air. Also note that privacy (sight and sound) should be maintained. A line of grilles specifically designed for this application is available from many manufacturers.

11-5 Filter Grilles

In general, the velocity through filters should be fairly low because filter performance and pressure drop are quite sensitive to flow rate. Therefore, filter grilles should be sized for ether a face velocity of 200 fpm or a core velocity of 300 fpm.

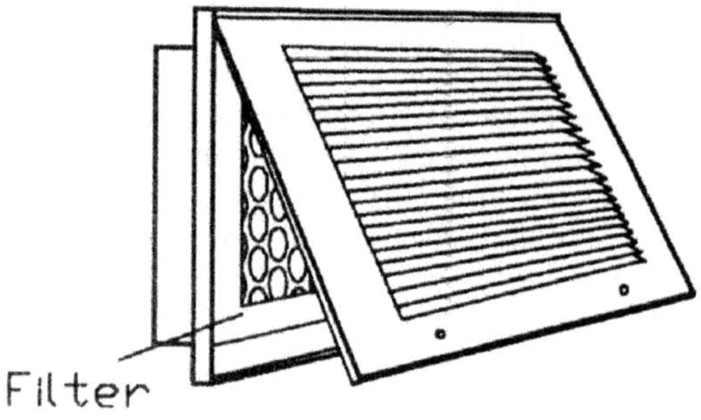

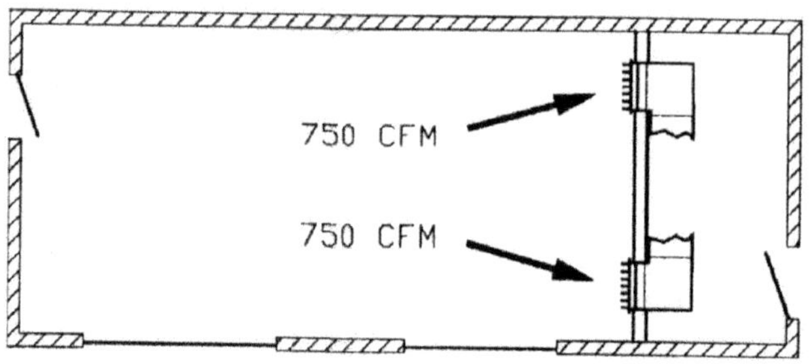

11-6 Residential Example

Size a standard residential grille and a residential filter grille to handle 1000 CFM of return air.

Solution - standard grille:

Standard grille gross area = 1000 / 300 = 3.33 Sq.Ft.

Approximate size (30 inches x 16 inches)

Solution - filter grille:

Filter grille gross area = 1000 / 200 = 5.00 Sq.Ft.

Approximately size (32 inches x 22 inches)

11-7 Commercial Example

The figure below shows two sidewall grilles could be used to return 1500 CFM of supply air. Refer to the NC table in Section 16-6 and to the return inlet data in Appendix A1 and use this information to size the returns.

The following table summarizes the performance of a few inlet grille sizes that qualify. These selections were based on two inlets returning 750 CFM each with an allowable NC rating of 35 or less. Of the various possible choices the 16x14 size or the 16x16 size are preferred because they have lower static pressure losses and lower face velocities.

No	CFM	Size	NC	Vel	sp
2	750	14x12	23	700	.089
2	750	14x14	20	650	.076
2	750	16x14	15	550	.055
2	750	16x16	8	460	.038

11-8 Transfer Grille Example

Size a transfer grille for a room that has 350 CFM of air supplied to it and a door undercut for a room that has 120 CFM of air supplied to it.

Solution - transfer grille:

Face area = 350 CFM / 200 fpm = 1.75 Sq.Ft.

Approximate size (12 inches x 22 inches)

Solution - door undercut:

Door undercut = 120 / 60 = 2 inches

Section 12

Balancing and Measuring Air Flow

Theoretically the residential duct sizing procedure produces a system that is somewhat balanced. In actual practice this may not be the case because the duct system design is based on a number of assumptions. In addition, sizing calculations and selections are based on maximum, minimum or average values and ultimately the system is assembled from standard sizes. As a result, each and every residential system must be adjusted in the field to assure that the proper air flow is being delivered to each space.

Low velocity commercial air distribution systems are usually designed by the "Equal Friction Method". This procedure produces a design that is inhearently unbalanced. In this case acceptable performance is only made possible by measuring and balancing the flow through the various outlets and inlets.

12.1 Balancing Procedures

In residential applications, temperature rise measurements across a heat exchanger or a DX coil provide information on the approximate air flow delivered to the system. Thermometers can also be used (on a design day) to adjust branch dampers and/or register dampers until the desired temperature in each room is achieved. A small air flow meter (velometer) and a simple cone shaped collection stack provide a quick and accurate method of measuring the air flow through a outlet.

For commercial systems, special air velocity meters, manometers, pitot tubes, etc., are required to measure flows and pressures at the outlets and inlets as well as at various points in the duct system. In addition, RPM counters, amprage meters, voltage meters etc. are required to test motor driven equipment performance. Furthermore, different types of HVAC systems require different testing procedures and reporting formats.

A discussion of the testing and balancing procedures that apply to either residential or commercial work is beyond the scope of this manual. Refer to other manuals which are exclusively devoted to describing the various testing and balancing procedures that apply to residential and commercial work.

12.2 Area Factor

Some manufacturers publish area factor (Ak) values which can be used to convert outlet velocity measurements into a corresponding flow rate. Ak values are only required when certain types of flowmeters and probes are used to measure the velocities between the outlet's vanes at certain positions around the outlet. (Note that there are different sets of Ak factors for different types of air velocity meters.) In the field, the published area factors for each outlet are multiplied by the average of the measured velocities to determine the CFM. The following equation applies:

CFM = Ak x Average Velocity

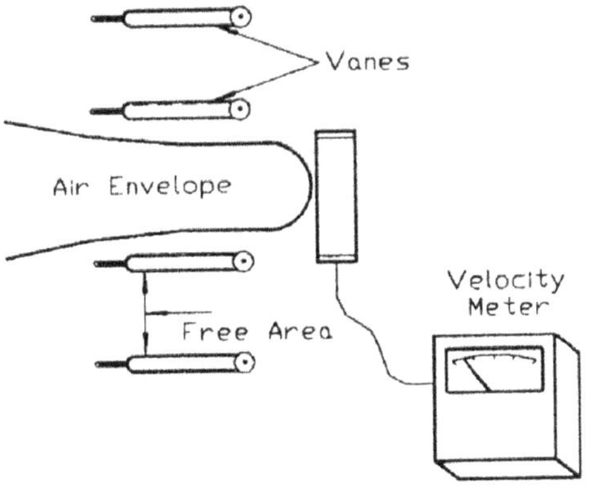

Appendix 1

Manufacturer's Application Data

Each manufacturer that publishes comprehensive air terminal data does so in a slightly different format. The following pages provide some examples of **Hart and Cooley** catalog data and **Titus Products** catalog data.

This data is also used as reference data for the example problems found in Sections 10 and 11. In actual practice it is necessary to have complete sets (of various manufacturers) catalog data on hand.

The data in this Appendix is used with the permission of Hart and Cooley, Inc., and Titus Products, Division of Philips Industries.

411 Deluxe Floor Diffuser

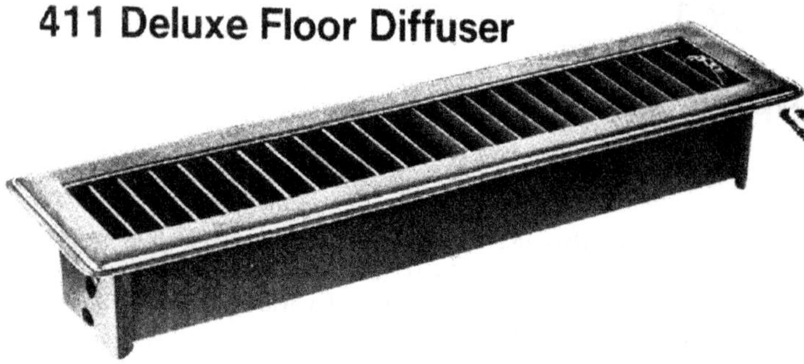

- ☐ All-steel construction
- ☐ Heavy-gauge individually welded diffusion vanes
- ☐ Multi-angle fin setting
- ☐ Foot-operated dial control
- ☐ Golden Sand or Dover White enamel finish

Only available in sizes shown

411 Floor Diffuser

Face Velocity		300	400	500	600	700	800	900	1000
Pressure Loss		.006	.010	.016	.022	.031	.040	.050	.062
2¼ x 10 Ak .09	cfm	25	35	45	55	60	70	80	90
	Spread	3	4	5	6.5	7.5	9	10	11
	Throw	2	2.5	3	4	4.5	5	6	6.5
2¼ x 12 Ak .11	cfm	35	45	55	65	75	90	100	110
	Spread	3	4.5	6	7	8.5	9.5	11	12
	Throw	2	3	3.5	4	5	6	6.5	7
2¼ x 14 Ak .13	cfm	40	50	65	80	90	105	115	130
	Spread	3.5	5	7	8	9.5	10.5	12	13
	Throw	2.5	3	4	4.5	5.5	6.5	7	8
4 x 10 Ak .17	cfm	50	70	85	100	120	135	155	170
	Spread	4	6	7.5	9	11	12	13.5	15.5
	Throw	3	3.5	4.5	5.5	6.5	7	8	9
4 x 12 Ak .21	cfm	65	85	105	125	145	170	190	210
	Spread	5	7	8.5	10	12	13.5	15.5	17
	Throw	3	4	5	6	7	8	9	10
4 x 14 Ak .25	cfm	75	100	125	150	175	200	225	250
	Spread	5.5	7.5	9.5	11	13	15	17	18.5
	Throw	3.5	4.5	5.5	6.5	7.5	9	10	11
6 x 10 Ak .27	cfm	80	110	135	160	190	215	245	270
	Spread	6	8	10	11.5	13.5	15.5	17.5	19.5
	Throw	3.5	4.5	6	7	8	9	10.5	11.5
6 x 12 Ak .33	cfm	100	130	165	200	230	265	295	330
	Spread	6.5	8.5	11	13	15	17	19.5	21.5
	Throw	4	5	6.5	7.5	9	10	11.5	12.5
6 x 14 Ak .39	cfm	115	155	195	235	275	310	350	390
	Spread	7.5	9.5	12	14	17	19	21.5	23.5
	Throw	4.5	5.5	7	8.5	10	11	12.5	14

Terminal velocity of 50 fpm.

406 Baseboard Diffuser

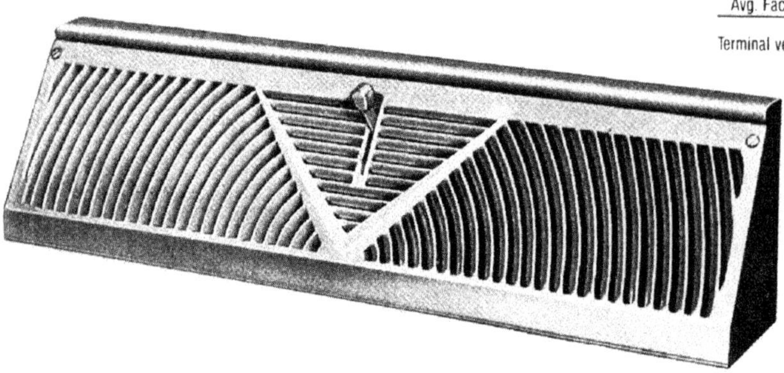

421 Floor Diffuser

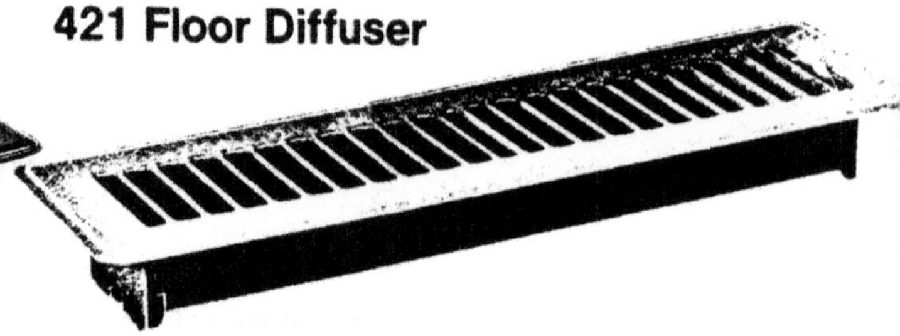

- ☐ All-steel construction
- ☐ Heavy-gauge stamped face
- ☐ Multi-angle fin setting
- ☐ Foot-operated dial control
- ☐ Golden Sand or Dover White enamel finish

Only available in sizes shown

421 Floor Diffuser

Face Velocity		300	400	500	600	700	800	900	1000
Pressure Loss		.006	.010	.016	.022	.031	.040	.050	.062
2¼ x 10 Ak .085	cfm		35	45	50	60	70	75	85
	Spread		4	5	6	7	8	9	10
	Throw		3	4.5	5	6	7	8	9
2¼ x 12 Ak .100	cfm	30	40	50	60	70	80	90	100
	Spread	3.5	4.5	5.5	7	8	9	10	11
	Throw	3	4	4.5	5.5	6.5	7	8	9
2¼ x 14 Ak .115	cfm	35	45	60	70	80	90	105	115
	Spread	3.5	4.5	6	7	8	9.5	10.5	12
	Throw	3.5	4	5	6	7	8	9	10
4 x 10 Ak .17	cfm	50	70	85	100	120	135	155	170
	Spread	4.5	6	8	10	11	12.5	14	15.5
	Throw	4	5	6.5	7.5	9	10	11.5	13
4 x 12 Ak .195	cfm	60	80	100	120	140	160	175	195
	Spread	5	6.5	8	9.5	11.5	13	14.5	16
	Throw	4	5.5	7	8	9.5	11	12	13
4 x 14 Ak .23	cfm	70	90	115	140	160	185	205	230
	Spread	5.5	7	8.5	10	12	13.5	15.5	17
	Throw	4.5	5.5	7	8.5	10	11.5	12.5	14

Terminal velocity of 50 fpm.

406 Baseboard Diffuser

	Boot Size	CFM/Cubic Feet per Minute						
		50	75	100	125	150	175	200
Pressure Loss (In W.G)	2¼ x 14	.005	.011	.020	.032	.045	.064	.082
	2¼ x 12	.006	.014	.025	.040	.056	.078	.100
Horiz. Spread (ft.)		3.5	4.5	6	7.5	8.5	9.5	10.5
Vertical Throw (ft.)		3.5	4.5	6.5	8	9	9.5	10.5
Avg. Face Velocity (fpm)		190	285	380	475	570	665	760

Terminal velocity of 50 fpm.

531 Royalaire® Floor Diffuser
Aluminum Face

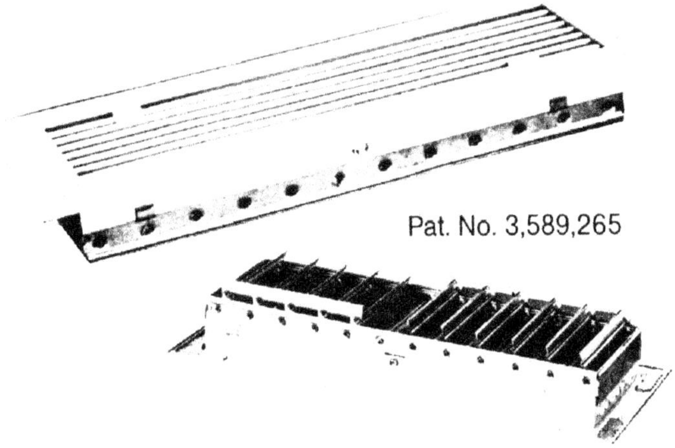

Pat. No. 3,589,265

- ☐ Extruded aluminum margins
- ☐ Heavy cross bracing at strategic intervals
- ☐ Dual multi-shutter valves for adjustable air pattern
- ☐ Nylon slide valve operators
- ☐ Clear anodized finish

Only available in sizes shown

531 Floor Diffuser

Face Velocity		300		400		500		600		700		800		900		1000	
Pressure Loss		.006		.010		.016		.022		.031		.040		.050		.062	
		H	C	H	C	H	C	H	C	H	C	H	C	H	C	H	C
2¼ x 10	cfm							35	35	40	40	45	50	50	55	55	60
Ak .055 H	Spread							3.5		4.5		5		5.5		5.5	
Ak .060 C	Throw							4	6	5.5	7.5	6	8.5	6.5	9	7	10
2¼ x 12	cfm					35	35	40	45	45	50	55	60	60	65	65	75
Ak .067 H	Spread					3.5		4		4.5		5		5.5		6*	
Ak .074 C	Throw					4	6	5	6.5	5.5	7.5	6	9	7	10	8	11
2¼ x 14	cfm			30	35	40	45	45	50	55	60	65	70	70	80	80	85
Ak .079 H	Spread			3		4		4.5		5		5.5		6*		6*	
Ak .087 C	Throw			3.5	4.5	5	6	5.5	7	6.5	8.5	7.5	10	8	11.5	8.5	12.5
4 x 10	cfm	35	40	45	50	55	65	70	75	80	90	90	100	105	115	115	125
Ak .115 H	Spread	3		3.5		4.5		5.5		6*		6*		6*		6*	
Ak .125 C	Throw	3	4.5	4	6	5.5	7.5	6.5	9.5	7.5	10.5	8.5	11.5	9.5	13.5	11.5	15
4 x 12	cfm	40	50	55	65	70	80	85	95	100	110	115	130	125	145	140	160
Ak .14 H	Spread	3		4		5		5.5		6*		6*		6*		6*	
Ak .16 C	Throw	3.5	5	4.5	6.5	6	8.5	7.5	10	9	11.5	10	13	11	15	12	17
4 x 14	cfm	50	55	65	70	80	90	100	110	115	125	130	145	150	160	165	180
Ak .165 H	Spread	3.5		4.5		5		6*		6*		6*		6*		6*	
Ak .18 C	Throw	4	5.5	5.5	7	6.5	9.5	8	11	10	12.5	11	14	12	15.5	13.5	17.5
6 x 10	cfm	60	60	75	80	95	105	115	125	135	145	150	165	170	185	190	205
Ak .19 H	Spread	3.5		4.5		5.5		6*		6*		6*		6*		6*	
Ak .205 C	Throw	4.5	6	5.5	7.5	7	9.5	8.5	11	10	13	11	15	12.5	16.5	14	18.5
6 x 12	cfm	70	75	90	100	115	125	135	150	160	175	180	200	205	225	225	250
Ak .225 H	Spread	4		5		6*		6*		6*		6*		6*		6*	
Ak .25 C	Throw	5	6.5	6	8.5	8	10.5	9	12	11	15	12.5	16.5	14	18.5	16	21
6 x 14	cfm	80	90	105	120	135	150	160	180	185	210	210	240	240	270	265	300
Ak .265 H	Spread	4.5		5		6*		6*		6*		6*		6*		6*	
Ak .30 C	Throw	5	7	7	9.5	8.5	11.5	10	13.5	12	16	13.5	18	16	20.5	17.5	23.5

The spread shown for the heating setting is for a valve setting of 22° right and left. *The maximum value given for spread for heating is that which occurs at the ceiling height (8 ft.) The cooling spread is a straight vertical column of air and is not shown. Terminal velocity of 50 fpm

462/464 Series Baseboard Diffusers

Diffuser Length Free Area		CFM (Cubic Feet per Minute)							
		50	70	90	110	130	150	170	190
	PL	.008	.015	.025	.038	.054	.072		
2 Feet 27 sq. in.	Spread	6	7	7.5	8.5	9	11		
	Throw	6	7.5	9	10.5	12	13		
	Velocity	267	373	480	587	693	800		
	PL	.006	.012	.021	.031	.043	.058	.072	.094
4 Feet 54 sq. in.	Spread	6	8	10	12	13.5	15	16.5	18
	Throw	4.5	5.5	6.5	7.5	8	9	9.5	10.5
	Velocity	153	186	240	293	347	400	453	503

Terminal velocity of 50 fpm

- ☐ All-steel construction
- ☐ Airfoil valves reduce resistance
- ☐ Equal air distribution over the length of the diffuser
- ☐ Dual valves on 464
- ☐ Face can be easily removed
- ☐ Dover White or Golden Sand enamel finish

464 Baseboard Diffuser
Four-foot length

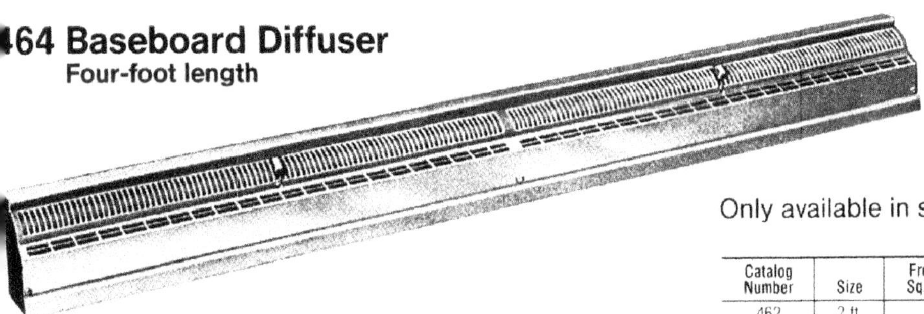

Only available in sizes shown

Catalog Number	Size	Free Area Sq. Inches	Overall Length	Height	Base Projection	Bottom Opening
462	2 ft	27	24"	4½"	2¾"	2⁵⁄₁₆" x 16"
464	4 ft	54	48"	4½"	2¾"	2⁵⁄₁₆" x 24"

TITUS® Round Diffusers • Description • Dimensions

Round Ceiling Diffusers

Model TMR • Steel
2 Horizontal Discharge Patterns

Model TMR-AA • Aluminum
2 Horizontal Discharge Patterns

- Model TMR and TMR-AA round ceiling diffusers are designed for both heating and cooling applications.
- All sizes have 3 cones, giving a uniform appearance where different sizes are used in the same area.
- Uniform, 360 degree discharge pattern.
- Excellent performance in variable air volume systems.
- The 2 horizontal discharge settings allow the diffuser to be adjusted for 2 different flow rate conditions. At Position 1, maximum capacity is obtained. At position 2, induction is increased, and the noise level is increased by 1 NC above those shown in the performance table on Page 39.
- Adjusted by removing the 2 inner cones as a unit and repositioning 6 screws.
- Spring lock allows easy removal and replacement of the 2 inner cones.
- Optional safety chain prevents dropping the inner cones after removal.
- Gasket around the outer edge guards against leakage and ceiling smudging.
- Material is steel or aluminum.

Standard Finish:

#25 White.

Optional Finish:

#01 Aluminum (Aluminum Diffusers Only).

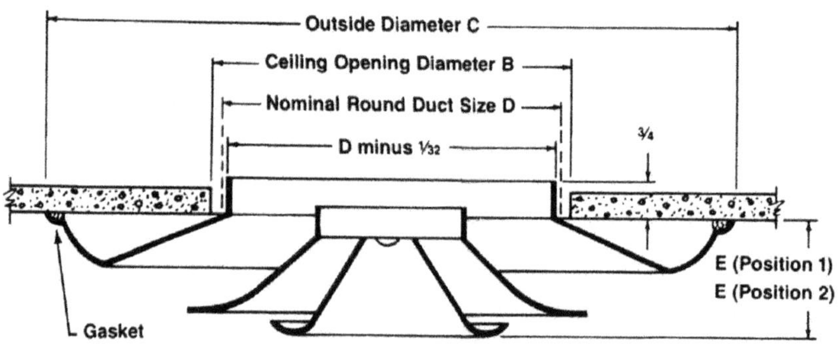

Nominal Round Duct Size D	B	C	E (Pos. 1)	E (Pos. 2)
6	6½	11⅛	1¾	1⅛
8	8½	14¾	2⅛	1½
10	10½	18¼	2⅞	2⅛
12	12½	22	3⅛	2⅜
14	14½	26	3⅜	2⅝
16	16½	29	4	3¼
18	18½	32½	4¾	3⅞
20†	20½	36	5⅞	4⅞
24†	24½	43¼	7¾	6⅝
30†	30½	53¾	8⅛	6⅝
36†	36½	64½	10⅛	8⅜

Dimensions are in inches.
†Sizes 20 through 36 are available in steel only.

TITUS® Round Diffusers • Performance

Models TMR, TMR-AA

Nominal Neck Size	Neck Velocity, FPM / Velocity Pressure, Inches WG	400 / .010	500 / .016	600 / .023	700 / .031	800 / .040	900 / .051	1000 / .063	1200 / .090	1400 / .122	1600 / .160
6" Dia.	Total Pressure, Inches WG	.026	.041	.059	.079	.102	.130	.161	.230	.311	.408
	Air Flow, CFM	**80**	**100**	**120**	**140**	**160**	**180**	**200**	**235**	**275**	**315**
	Radius of Diffusion, Feet	2-2-4	2-3-5	2-4-6	3-4-7	3-5-8	4-5-9	4-6-10	5-7-11	6-8-13	6-10-15
	NC (Noise Criterion)	—	—	14	19	23	26	30	35	40	44
8" Dia.	Total Pressure, Inches WG	.033	.052	.075	.101	.130	.166	.205	.292	.397	.520
	Air Flow, CFM	**140**	**175**	**210**	**245**	**280**	**315**	**350**	**420**	**490**	**560**
	Radius of Diffusion, Feet	2-4-6	3-4-7	4-5-9	4-6-10	5-7-11	5-8-13	6-9-14	7-11-17	8-13-20	10-14-23
	NC (Noise Criterion)	—	15	21	26	31	34	37	44	48	53
10" Dia.	Total Pressure, Inches WG	.027	.043	.062	.084	.108	.138	.170	.243	.329	.432
	Air Flow, CFM	**220**	**270**	**330**	**380**	**435**	**490**	**545**	**655**	**765**	**870**
	Radius of Diffusion, Feet	3-4-7	3-5-8	4-6-10	5-7-11	5-8-13	6-9-15	7-10-16	8-12-20	10-14-23	11-16-26
	NC (Noise Criterion)	—	11	17	21	26	30	33	39	44	48
12" Dia.	Total Pressure, Inches WG	.026	.042	.060	.081	.105	.134	.166	.236	.320	.420
	Air Flow, CFM	**315**	**390**	**470**	**550**	**630**	**705**	**785**	**940**	**1100**	**1255**
	Radius of Diffusion, Feet	3-5-8	4-6-10	5-7-12	6-8-13	6-10-15	7-11-17	8-12-19	10-14-23	11-17-27	13-19-31
	NC (Noise Criterion)	—	11	17	22	26	30	33	39	45	49
14" Dia.	Total Pressure, Inches WG	.038	.061	.087	.118	.152	.194	.240	.342	.465	.608
	Air Flow, CFM	**425**	**530**	**635**	**745**	**850**	**955**	**1060**	**1270**	**1490**	**1695**
	Radius of Diffusion, Feet	4-6-10	5-8-12	6-9-15	7-11-17	8-12-20	9-14-22	10-15-24	12-18-29	14-21-34	16-24-39
	NC (Noise Criterion)	11	18	23	28	32	36	40	46	51	55
16" Dia.	Total Pressure, Inches WG	.033	.053	.076	.102	.132	.168	.208	.297	.403	.528
	Air Flow, CFM	**560**	**700**	**840**	**980**	**1120**	**1260**	**1400**	**1680**	**1960**	**2240**
	Radius of Diffusion, Feet	5-7-11	6-9-14	7-10-16	8-12-19	9-14-22	10-15-24	11-17-27	14-20-33	16-24-38	18-27-43
	NC (Noise Criterion)	—	16	22	27	31	35	39	44	49	53
18" Dia.	Total Pressure, Inches WG	.030	.048	.069	.093	.120	.153	.189	.270	.366	.480
	Air Flow, CFM	**710**	**885**	**1060**	**1240**	**1420**	**1590**	**1770**	**2120**	**2480**	**2830**
	Radius of Diffusion, Feet	5-7-12	6-9-15	7-11-18	9-13-21	10-15-24	11-17-27	12-19-30	15-22-36	17-26-42	20-30-48
	NC (Noise Criterion)	—	15	21	26	30	34	37	43	48	52
20" Dia.	Total Pressure, Inches WG	.030	.047	.068	.091	.118	.150	.186	.266	.360	.472
	Air Flow, CFM	**875**	**1100**	**1310**	**1530**	**1750**	**1970**	**2190**	**2610**	**3060**	**3500**
	Radius of Diffusion, Feet	6-8-13	7-10-17	8-12-20	10-14-23	11-17-27	12-19-30	14-21-33	16-25-40	19-29-46	22-33-53
	NC (Noise Criterion)	—	16	21	26	30	34	38	43	48	52
24" Dia.	Total Pressure, Inches WG	.024	.038	.054	.073	.094	.120	.148	.211	.286	.376
	Air Flow, CFM	**1260**	**1570**	**1880**	**2200**	**2510**	**2820**	**3140**	**3770**	**4400**	**5020**
	Radius of Diffusion, Feet	6-9-15	8-12-19	9-14-22	11-16-26	12-19-30	14-21-34	16-23-37	19-28-45	22-33-52	25-37-60
	NC (Noise Criterion)	—	13	19	24	28	32	35	41	46	50
30" Dia.	Total Pressure, Inches WG	.031	.050	.071	.096	.124	.158	.195	.279	.378	.496
	Air Flow, CFM	**1960**	**2450**	**2940**	**3430**	**3920**	**4410**	**4900**	**5880**	**6860**	**7840**
	Radius of Diffusion, Feet	8-13-20	10-16-25	13-19-30	15-22-35	17-25-40	19-28-45	21-31-50	25-38-60	29-44-70	33-50-80
	NC (Noise Criterion)	11	18	24	26	30	34	41	47	51	56
36" Dia.	Total Pressure, Inches WG	.024	.038	.055	.074	.096	.122	.151	.216	.293	.384
	Air Flow, CFM	**2820**	**3520**	**4230**	**4930**	**5630**	**6340**	**7040**	**8450**	**9850**	**11,260**
	Radius of Diffusion, Feet	9-14-22	12-18-28	14-21-34	16-25-39	19-28-45	21-32-50	23-35-56	28-42-67	33-49-78	37-56-90
	NC (Noise Criterion)	—	15	21	26	30	34	38	43	48	53

- All pressures are in inches of water
- Radius of diffusion values shown above are for Positions 1 and 2.
- To obtain the static pressure, subtract the velocity pressure from the total pressure. Radius of diffusion values are given for terminal velocities of 150, 100, and 50 fpm.
- If the diffuser is mounted on an exposed duct, the radius of diffusion values are 0.70 times those listed in the table.
- NC values are based on a room absorption of 10 dB, re 10^{-12} watts, for the full-horizontal discharge setting. Add 1 NC for Position 2.
- Diffusers are shipped in Position 1.
- Dash (—) in space denotes an NC value less than 10.
- Data were obtained from tests conducted in accordance with ISO Standard 5219, ISO Standard 3741 and ADC Test Code 1062 GRD84.

TITUS® Square Ceiling Diffusers • Description

Square Ceiling Diffusers High Performance Round Neck

Model TMS • Steel
Model TMS-AA • Aluminum

• TITUS® Models TMS and TMS-AA square ceiling diffusers deliver supply air in a true 360° pattern. Designed to protect ceilings from streaking and smudging.

• Excellent performance in variable air volume systems. The uniform, nearly horizontal jet from the outer cone maintains effective room air distribution even when the air volume varies over a considerable range.

• All sizes have 3 cones, giving a uniform appearance where different sizes are used in the same area.

• Screwdriver adjustment of the optional damper without removing the inner cone. (See Diagram 2 on the facing page.)

• Quick removal of the inner cone with a hex key. (See Diagram 3 on the facing page.)

• All cones are die-stamped, one-piece construction. Smooth, clean surfaces with no corner joints.

• Material is heavy gauge steel or aluminum, according to the model selected.

Standard Finish:

#25 White.

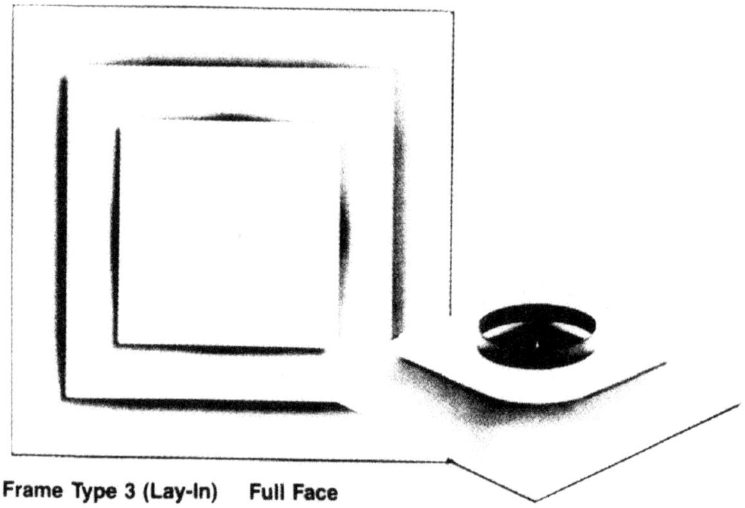

Frame Type 3 (Lay-In) Full Face

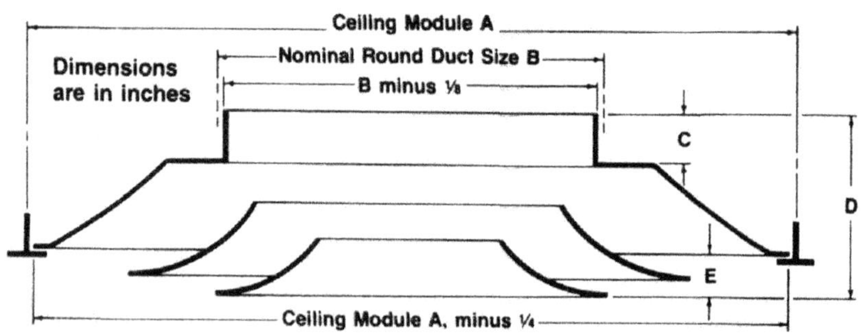

Ceiling Module A	Nominal Round Duct Size B	C	D	E
12 × 12	4,5*	3⅛	5⅜	1⅛
12 × 12	6,7	1⅛	3⅜	1⅛
12 × 12	8	1¼	3½	1⅛
24 × 24	8	1¼	5	1¼
24 × 24	10,12,14,15	1⅜	5⅛	1¼

* Adapter is provided for Sizes 4 and 5.

Frame Type 3 (Lay-In) Panel Mounted

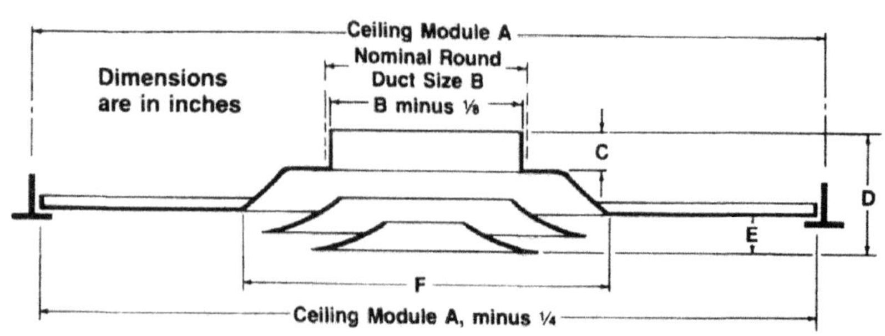

Ceiling Module A	Nominal Round Duct Size B	Nominal Face Size F	C	D	E
24 × 24	4,5*	12 × 12	3⅛	5⅜	1⅛
24 × 24	6,7	12 × 12	1⅛	3⅜	1⅛
24 × 24	8	12 × 12	1¼	3½	1⅛

* Adapter is provided for Sizes 4 and 5.

TITUS[R] Square Ceiling Diffusers • Performance

odels TMS, TMS-AA • 12 × 12 Face Size

NC-20

ominal Neck Size	Neck Velocity, FPM / Velocity Pressure, Inches WG	400 .010	500 .016	600 .023	700 .031	800 .040	900 .051	1000 .063	1200 .090	1400 .122	1600 .160
4" Dia.	Total Pressure, Inches WG	.011	.017	.024	.033	.043	.055	.067	.097	.132	.172
	Air Flow, CFM	**35**	**44**	**52**	**61**	**70**	**79**	**87**	**105**	**122**	**140**
	Radius of Diffusion, Feet	·1-1-2	1-1-3	1-2-4	1-2-4	2-2-5	2-3-5	2-3-6	2-4-6	3-4-7	3-5-7
	NC (Noise Criterion)	—	—	—	—	—	11	14	20	25	29
5" Dia.	Total Pressure, Inches WG	.011	.017	.024	.033	.043	.035	.067	.097	.132	.172
	Air Flow, CFM	**54**	**68**	**82**	**95**	**109**	**122**	**136**	**163**	**190**	**218**
	Radius of Diffusion, Feet	1-1-3	1-2-4	1-2-4	2-3-5	2-3-6	2-3-7	2-4-7	3-4-8	3-5-8	4-6-9
	NC (Noise Criterion)	—	—	—	—	—	11	14	20	25	29
6" Dia.	Total Pressure, Inches WG	.019	.029	.042	.057	.075	.095	.117	.169	.230	.300
	Air Flow, CFM	**78**	**98**	**118**	**137**	**157**	**176**	**196**	**235**	**274**	**314**
	Radius of Diffusion, Feet	2-2-5	2-3-6	2-4-7	3-4-8	3-5-9	4-5-10	4-6-10	5-7-11	5-9-12	6-7-13
	NC (Noise Criterion)	—	—	11	16	20	24	22	34	39	43
7" Dia.	Total Pressure, Inches WG	.021	.033	.048	.065	.085	.107	.132	.190	.259	.338
	Air Flow, CFM	**107**	**134**	**160**	**187**	**214**	**240**	**267**	**320**	**374**	**427**
	Radius of Diffusion, Feet	2-3-5	2-3-7	3-4-8	3-5-10	4-5-10	4-6-11	5-7-12	5-8-13	6-10-14	7-10-15
	NC (Noise Criterion)	—	—	11	16	20	24	28	34	39	43
8" Dia.	Total Pressure, Inches WG	024	.037	.054	.073	.096	.121	.150	.215	.273	.383
	Air Flow, CFM	**140**	**175**	**209**	**244**	**279**	**314**	**349**	**419**	**489**	**558**
	Radius of Diffusion, Feet	2-3-6	3-4-8	3-5-9	4-5-11	4-6-12	5-7-13	5-8-13	6-9-15	7-11-16	8-12-17
	NC (Noise Criterion)	—	—	11	16	20	24	28	34	39	43

NC-20 NC-30 NC-40

odels TMS, TMS-AA • 24 × 24 Face Size

NC-20 NC-30

ominal Neck Size	Neck Velocity, FPM / Velocity Pressure, Inches WG	400 .010	500 .016	600 .023	700 .031	800 .040	900 .051	1000 .063	1200 .090	1400 .122	1600 .160
8" Dia.	Total Pressure, Inches WG	.016	.025	.036	.049	.064	.081	.100	.145	.197	.257
	Air Flow, CFM	**140**	**175**	**200**	**244**	**279**	**314**	**349**	**419**	**489**	**558**
	Radius of Diffusion, Feet	2-3-6	2-4-7	3-4-9	3-5-10	4-6-11	4-6-12	5-7-13	6-9-14	7-10-15	8-11-16
	NC (Noise Criterion)	—	—	13	17	21	24	27	32	37	40
10" Dia.	Total Pressure, Inches WG	.016	.026	.037	.050	.066	.023	.103	.148	.202	.263
	Air Flow, CFM	**218**	**273**	**327**	**382**	**436**	**490**	**545**	**654**	**763**	**872**
	Radius of Diffusion, Feet	2-4-7	3-4-9	4-5-11	4-6-12	5-7-14	5-8-15	6-9-16	7-11-17	8-12-19	9-14-20
	NC (Noise Criterion)	—	11	16	20	24	27	30	35	40	44
12" Dia.	Total Pressure, Inches WG	.017	.026	.038	.052	.068	.086	.106	.153	.208	.271
	Air Flow, CFM	**314**	**393**	**471**	**550**	**628**	**707**	**785**	**842**	**1099**	**1256**
	Radius of Diffusion, Feet	3-4-9	4-5-11	4-9-13	5-7-15	6-7-17	6-10-18	7-11-19	9-13-21	10-15-22	11-17-24
	NC (Noise Criterion)	—	13	19	23	27	30	33	38	42	46
14" Dia.	Total Pressure, Inches WG	.018	.027	.040	.054	.070	.089	.110	.158	.215	.281
	Air Flow, CFM	**428**	**535**	**641**	**748**	**855**	**962**	**1069**	**1283**	**1497**	**1710**
	Radius of Diffusion, Feet	3-5-10	4-6-12	5-7-15	6-9-17	7-10-20	7-11-21	8-12-22	10-15-24	12-17-26	13-20-28
	NC (Noise Criterion)	—	16	21	25	29	32	35	40	44	48
15" Dia.	Total Pressure, Inches WG	.018	.028	.040	.055	.072	.091	.112	.162	.220	.287
	Air Flow, CFM	**491**	**614**	**736**	**859**	**982**	**1104**	**1227**	**1472**	**1718**	**1963**
	Radius of Diffusion, Feet	4-5-11	4-7-13	5-8-16	6-9-19	7-11-21	8-12-22	9-13-24	11-16-26	12-19-28	11-21-30
	NC (Noise Criterion)	10	16	22	26	30	33	36	41	45	49

NC-40

All pressures are in inches of water.
Radius of diffusion values are given r terminal velocities of 150, 100 and) fpm.
NC values are based on a room ɔsorption of 10 dB, re 10^{-12} watts.

• Shaded dividing lines denote ranges of NC values.
• Dash (—) in space denotes NC value less than 10.

• Data were obtained from tests conducted in accordance with ISO Standard 5219, ISO Standard 3741 and ADC Test Code 1062 GRD84.

301 & A301 Register
One-way deflection

302 & A302 Register
Two-way deflection

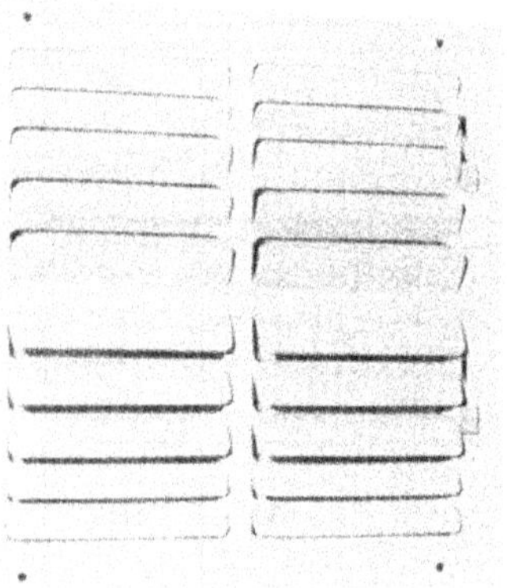

303 & A303 Register
Three-way deflection

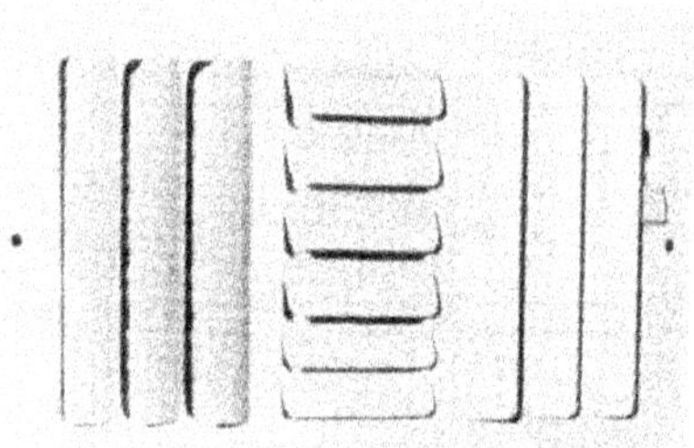

304 & A304 Register
Four-way deflection

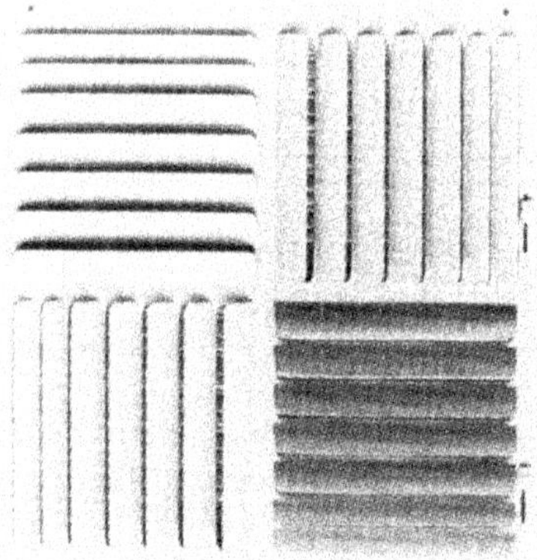

A300 Series Curved Blade Diffusers

Face Velocity		300	400	500	600	700	800	900	1000	1100
Pressure Loss		.006	.010	.016	.022	.031	.040	.050	.062	.075
6 x 6	cfm	25	35	45	55	60	70	80	90	100
	Throw 301	4.5	6.5	8.5	10	11	13	15	17	19
Ak .088	Throw 302	3.5	5.0	6.0	7.5	8.5	9.5	11	12	14
	Throw 303	2.5	3.5	4.5	5.5	6.0	7.0	8.0	8.5	9.5
	Throw 304	2.5	3.5	4.5	5.5	6.0	7.0	8.0	9.0	10
8 x 4	cfm	25	30	40	45	55	60	70	80	85
	Throw 301	5.0	6.5	8.0	9.5	11	13	14	16	18
Ak .078	Throw 302	3.5	4.5	6.0	7.0	8.0	9.5	11	12	13
	Throw 303	3.0	3.5	4.5	5.5	6.5	7.5	8.5	9.5	10
8 x 6	cfm	35	45	60	70	85	95	105	120	130
	Throw 301	6.0	8.0	10	12	14	16	17	19	21
Ak .118	Throw 302	4.0	5.5	7.0	8.5	10	11	13	14	15
	Throw 303	3.5	4.5	5.5	6.5	8.0	9.0	10	11	12
8 x 8	cfm	45	65	80	95	110	125	140	160	175
	Throw 301	6.5	9.0	11	13	16	18	20	22	25
Ak .158	Throw 302	4.5	6.5	8.0	9.5	11	13	14	16	17
	Throw 303	4.0	5.0	6.5	7.5	9.0	10	12	13	14
	Throw 304	3.5	4.5	6.0	7.0	8.0	9.0	10	12	13
10 x 4	cfm	30	40	50	60	70	80	90	100	110
	Throw 301	5.5	7.0	9.0	11	13	15	16	18	20
Ak .098	Throw 302	4.0	5.5	7.0	8.5	10	11	13	14	15
	Throw 303	3.0	4.0	5.0	6.0	7.0	8.0	9.0	10	11
10 x 6	cfm	45	60	75	90	105	120	135	150	165
	Throw 301	6.5	8.5	11	13	15	17	20	22	24
Ak .149	Throw 302	5.0	6.5	8.0	9.5	11	13	15	16	18
	Throw 303	3.5	5.0	6.0	7.5	8.5	9.5	11	12	13
10 x 8	cfm	60	80	100	120	140	160	180	195	215
	Throw 301	7.5	10	12	15	17	20	22	24	27
Ak .197	Throw 302	5.5	7.5	9.0	11	13	15	16	18	20
	Throw 303	4.0	5.5	7.0	8.5	9.5	11	13	14	15
10 x 10	cfm	75	100	125	150	175	200	225	250	280
	Throw 301	8.5	11	14	17	20	23	26	28	32
Ak .252	Throw 302	6.0	8.0	10	12	14	16	18	20	23
	Throw 303	5.0	6.5	8.0	9.5	11	13	14	16	18
	Throw 304	4.5	5.5	7.0	8.5	10	11	13	14	16
12 x 4	cfm	35	45	60	70	85	95	105	120	130
	Throw 301	6.0	8.0	10	12	14	16	17	20	21
Ak .118	Throw 302	4.5	6.0	7.5	9.0	11	12	13	15	16
	Throw 303	3.5	4.5	6.0	7.0	8.5	9.0	10	12	13
12 x 6	cfm	55	70	90	105	125	145	160	180	195
	Throw 301	7.0	9.0	12	14	16	19	21	24	26
Ak .179	Throw 302	5.5	7.0	9.0	10	12	14	16	18	19
	Throw 303	4.5	5.5	7.0	8.0	9.5	11	12	14	15
12 x 8	cfm	75	95	120	145	170	195	220	240	270
	Throw 301	8.5	11	14	16	19	22	25	27	30
Ak .242	Throw 302	6.0	8.0	10	12	14	16	18	20	22
	Throw 303	5.0	6.5	8.0	9.5	11	13	15	16	18
12 x 12	cfm	110	145	185	220	260	290	330	370	400
	Throw 301	10	13	17	20	24	27	31	34	37
Ak .367	Throw 302	7.0	9.5	12	14	17	19	22	24	26
	Throw 303	6.0	8.0	10	12	14	16	18	20	22
	Throw 304	5.0	6.5	8.5	10	12	13	15	17	19
14 x 4	cfm	40	55	70	85	100	110	120	140	150
	Throw 301	6.0	8.5	11	13	15	17	18	21	23
Ak .138	Throw 302	4.5	6.5	8.0	10	12	13	14	16	17
	Throw 303	3.5	5.0	6.0	7.5	8.5	9.5	10	12	13
14 x 6	cfm	65	85	105	125	145	170	190	210	230
	Throw 301	8.0	10	13	15	18	21	23	25	28
Ak .210	Throw 302	6.0	7.5	9.5	11	13	15	17	19	21
	Throw 303	4.5	6.0	7.5	8.5	10	12	13	15	16
14 x 8	cfm	85	110	140	165	195	220	250	280	310
	Throw 301	9.0	11	15	17	20	23	26	29	32
Ak .278	Throw 302	6.5	8.5	11	13	15	17	19	21	24
	Throw 303	5.0	6.5	8.5	10	12	13	16	17	19
14 x 14	cfm	150	200	250	300	350	400	460	510	560
	Throw 301	12	16	20	24	28	31	36	40	44
Ak .505	Throw 302	8.5	11	14	17	19	22	26	28	31
	Throw 303	7.0	9.0	11	14	16	18	21	23	25
	Throw 304	6.0	8.0	10	12	14	16	18	20	22
16 x 4	cfm	47	65	80	95	110	125	140	160	175
	Throw 301	6.5	9.0	11	13	16	18	20	22	25
Ak .158	Throw 302	4.5	6.5	8.0	9.5	11	13	14	16	17
	Throw 303	4.0	5.0	6.5	7.5	9.0	10	12	13	14
16 x 6	cfm	75	95	120	145	170	195	220	240	270
	Throw 301	8.5	11	14	16	19	22	25	27	30
Ak .242	Throw 302	6.0	8.0	10	12	14	16	18	20	22
	Throw 303	5.0	6.0	7.5	9.0	11	12	14	15	17
16 x 8	cfm	100	130	160	190	220	260	290	320	350
	Throw 301	9.5	13	15	18	21	25	28	31	34
Ak .320	Throw 302	7.0	9.0	11	13	16	18	20	23	25
	Throw 303	5.5	7.0	9.0	10	12	14	16	17	19
16 x 16	cfm	185	240	310	370	430	490	550	610	670
	Throw 301	13	16	21	25	29	33	38	42	46
Ak .611	Throw 302	9.0	12	15	18	21	24	27	29	32
	Throw 303	7.0	9.5	12	14	17	19	21	24	26
	Throw 304	6.5	8.0	11	13	15	17	19	21	23

Terminal velocity of 75 fpm

C3 Grille
CD3 Register, as above, with opposed blade damper
CM3 Register, as above, with multi-shutter valve

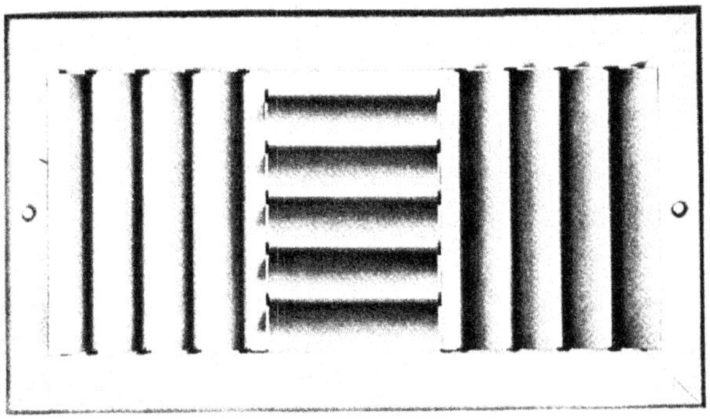

C3

C4 Grille
CD4 Register, as above, with opposed blade damper
CM4 Register, as above, with multi-shutter valve

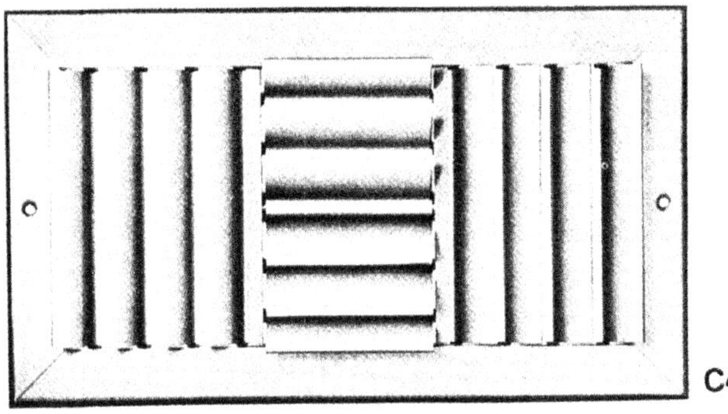

C4

: Series Curved Blade iffusers ne-Way, Two-Way

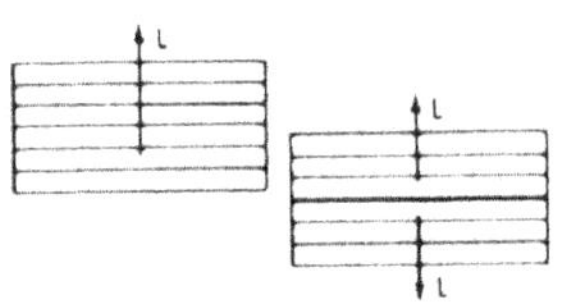

ce Velocity		400	500	600	700	800	900	1000	1100	1200
ssure Loss		.010	.016	.022	.031	.040	.050	.062	.075	.090
6 .09	cfm Throw 1/2	35 3.5/2.5	45 5/3.5	55 6/4	65 7/5	70 7.5/5.5	80 8.5/6	90 9.5/7	100 11/7.5	110 11.5/8.5
6 .10	cfm Throw 1/2	40 3.5/2.5	50 4.5/3	60 5.5/4	70 6.5/4.5	80 7/5	90 8/6	100 9/6.5	110 10/7	120 11/7.5
x 6 .15	cfm Throw 1/2	60 5/3.5	75 6/4.5	90 7/5	105 8.5/6	120 9.5/7	135 11/7.5	150 12/8.5	165 13/9.5	180 14/10
8 .16	cfm Throw 1/2	65 5/3.5	80 6/4.5	95 7.5/5	110 8.5/6	130 10/7	145 11/8	160 12/9	175 14/9.5	190 15/10
x 6 .18	cfm Throw 1/2	70 5/3.5	90 6.5/4.5	110 8/5.5	125 9/6.5	145 11/7.5	160 12/8.5	180 13/9.5	200 15/10	215 16/11
x 6 .21	cfm Throw 1/2	85 5.5/4	105 7/5	125 8.5/6	145 10/7	170 11/8	190 13/9	210 14/10	230 16/11	250 17/12
x 10 .24	cfm Throw 1/2	95 6/4	120 7.5/5	145 9/6.5	170 10/7.5	190 12/8	215 13/9.5	240 15/10	265 16/11	290 18/13
x 10 .29	cfm Throw 1/2	115 6.5/4.5	145 8/5.5	175 9.5/7	205 11/8	230 13/9	260 14/10	290 16/11	320 18/13	350 19/14
x 8 .31	cfm Throw 1/2	125 6.5/5	155 8.5/6	185 10/7	215 12/8	250 13/9.5	280 15/11	310 17/12	340 18/13	370 20/14
x 12 .35	cfm Throw 1/2	140 7/5	175 9/6	210 11/7.5	245 12/8.5	280 14/10	315 16/11	350 18/12	385 19/14	420 21/15
x 12 .46	cfm Throw 1/2	185 8/5.5	230 10/7	275 12/8.5	320 14/10	370 16/11	415 18/13	460 20/14	505 22/16	550 24/17
x 4 .48	cfm Throw 1/2	190 8/5.5	240 10/7.5	290 12/9	335 14/10	385 17/12	430 18/13	480 21/15	530 23/16	575 25/17
x 16 .63	cfm Throw 1/2	250 9.5/6.5	315 12/8.5	380 14/10	440 16/12	505 19/13	565 21/15	630 23/17	695 26/18	755 28/20
x 14 .68	cfm Throw 1/2	270 9.5/7	340 12/8.5	410 15/10	475 17/12	545 19/14	610 22/15	680 24/17	750 27/19	815 29/21
x 12 .70	cfm Throw 1/2	280 10/7	350 12/8.5	420 15/10	490 17/12	560 20/14	630 22/16	700 25/17	770 27/19	840 30/21
x 10 .73	cfm Throw 1/2	290 10/7	365 13/9	440 15/11	510 18/12	585 20/14	655 23/16	730 25/18	805 28/20	875 30/21
x 10 .88	cfm Throw 1/2	350 11/8	440 14/10	530 17/12	615 19/14	705 22/16	790 25/18	880 28/20	970 31/22	1055 33/24
x 12 1.05	cfm Throw 1/2	420 12/8.5	525 15/11	630 18/13	735 21/15	840 24/17	945 27/190	1050 30/21	1155 33/23	1260 36/25
x 16 1.15	cfm Throw 1/2	460 12/9	575 16/11	690 19/13	805 22/15	920 25/18	1035 28/20	1150 31/22	1265 34/24	1380 37/26
x 16 1.4	cfm Throw 1/2	560 14/9.5	700 17/12	840 21/15	980 24/17	1120 27/19	1260 31/22	1400 34/24	1540 38/27	1680 41/29

rminal velocity of 75 fpm.

C Series Four-Way

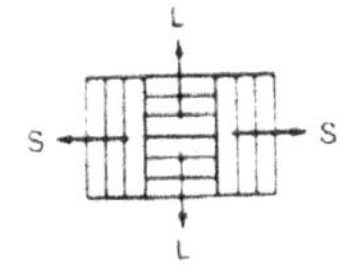

Face Velocity		400	500	600	700	800	900	1000	1100	1200
Pressure Loss		.010	.016	.022	.031	.040	.050	.062	.075	.090
6 x 6 Ak .09	Total cfm cfm L/S Throw L/S	35 5/13 1.5/2	45 6/17 1.5/3	55 7/20 2/3.5	65 9/24 2.5/4	70 9/26 2.5/4.5	80 11/29 3/5	90 12/33 3.5/6	100 13/37 4/6.5	110 15/40 4.5/7
8 x 6 Ak .10	Total cfm cfm L/S Throw L/S	40 9/11 1.5/1.5	50 12/13 2.5/2	60 14/16 2.5/2.5	70 16/19 3/2.5	80 18/22 3.5/3	90 21/24 4/3.5	100 23/27 4.5/4	110 25/30 4.5/4.5	120 28/32 5.5/4.5
10 x 6 Ak .15	Total cfm cfm L/S Throw L/S	60 11/19 2/2.5	75 14/24 2.5/3.5	90 16/29 3/4	105 19/33 3.5/4.5	120 22/38 4/5.5	135 25/43 4.5/6	150 27/48 5/7	165 30/52 5.5/7.5	180 33/57 6/8
8 x 8 Ak .16	Total cfm cfm L/S Throw L/S	65 15/17 2.5/2	80 18/22 3/2.5	95 22/26 3.5/3	110 25/30 4/3.5	130 30/35 5/4.5	145 33/39 3.5/5	160 37/43 6/5.5	175 40/47 6.5/6	190 44/51 7/6.5
12 x 6 Ak .18	Total cfm cfm L/S Throw L/S	70 10/25 2/3	90 13/32 2.5/4	110 16/39 3/5	125 19/44 3.5/5.5	145 22/51 4/6.5	160 24/56 4.5/7	180 27/63 5/8	200 30/70 5.5/8.5	215 32/75 6/9.5
14 x 6 Ak .21	Total cfm cfm L/S Throw L/S	85 11/32 2/3.5	105 13/39 2.5/4.5	125 16/47 3/5	145 18/54 3.5/6	170 22/63 4/7	190 24/71 4.5/8	210 27/78 5/8.5	230 29/86 5.5/9.5	250 32/93 6/10
10 x 10 Ak .24	Total cfm cfm L/S Throw L/S	95 17/30 2.5/3	120 22/38 3/4	145 26/46 3.5/5	170 31/54 4.5/6	190 35/60 5/6.5	215 39/68 5.5/7.5	240 44/76 6.5/8	265 48/84 7/9	290 53/92 7.5/10
12 x 10 Ak .29	Total cfm cfm L/S Throw L/S	115 17/40 2.5/4	145 22/51 3/5	175 26/61 3.5/5.5	205 31/72 4.5/7	230 35/80 5/7.5	250 39/91 5.5/8.5	290 44/101 6.5/9.5	320 48/112 7/11	350 53/122 7.5/11
16 x 8 Ak .31	Total cfm cfm L/S Throw L/S	125 22/41 3/4	155 27/50 3.5/4.5	185 33/60 4/5.5	215 38/70 5/6.5	250 44/81 5.5/7.5	280 49/91 6.5/8.5	310 55/100 7/9.5	340 60/110 7.5/10	370 65/120 8.5/11
12 x 12 Ak .35	Total cfm cfm L/S Throw L/S	140 21/49 2.5/4	175 26/61 3.5/5	210 31/74 4/6.5	245 37/86 5/7.5	280 42/98 5.5/8.5	315 47/110 6/9.5	350 52/123 7/10	385 58/135 7.5/11	420 63/147 8/13
16 x 12 Ak .46	Total cfm cfm L/S Throw L/S	185 33/60 3.5/4.5	230 40/75 4/5.5	275 48/89 5/7	320 56/104 6/8	370 65/120 7/9	415 73/135 7.5/10	460 81/149 8.5/11	505 89/164 9/12	550 97/178 10/14
14 x 14 Ak .48	Total cfm cfm L/S Throw L/S	190 24/71 3/5	240 31/89 3.5/6.5	290 37/108 4.5/7.5	335 43/125 5/9	385 49/143 6/10	430 55/160 6.5/11	480 61/179 7.5/13	530 68/197 8/14	575 74/214 9/15
16 x 16 Ak .63	Total cfm cfm L/S Throw L/S	250 44/81 4/5.5	315 55/102 5/6.5	380 67/123 6/8	440 77/143 7/9.5	505 89/164 8/11	565 99/183 9/12	630 111/204 10/13	695 122/225 11/15	755 133/245 12/16
20 x 14 Ak .68	Total cfm cfm L/S Throw L/S	270 38/97 3.5/6	340 48/122 4.5/7	410 57/148 5.5/9	475 67/171 6.5/10	545 76/196 7/12	610 85/220 8/13	680 95/245 9/15	750 105/270 10/16	815 114/293 11/17
24 x 12 Ak .70	Total cfm cfm L/S Throw L/S	280 45/95 4/5.5	350 56/119 5/7	420 67/143 6/8.5	490 78/167 7/10	560 89/191 8/12	630 100/215 9/13	700 111/239 10/14	770 122/263 11/16	840 134/286 12/17
30 x 10 Ak .73	Total cfm cfm L/S Throw L/S	290 46/99 4/6	365 58/124 5/7.5	440 70/150 6/9	510 82/173 7/10	585 94/199 8/12	655 105/223 9/13	730 117/248 10/15	805 129/274 11/16	875 140/296 12/18
36 x 10 Ak .88	Total cfm cfm L/S Throw L/S	350 57/118 4.5/6.5	440 71/149 5.5/8	530 86/179 6.5/9.5	615 100/208 8/11	705 114/238 9/13	790 128/267 10/14	880 143/297 11/16	970 157/328 12/18	1055 171/357 13/19
36 x 12 Ak 1.05	Total cfm cfm L/S Throw L/S	420 67/142 5/7	525 85/178 6/9	630 101/214 7/11	735 118/249 8.5/12	840 135/285 9.5/14	945 152/320 11/16	1050 169/356 12/18	1155 186/392 13/19	1260 203/427 14/21
30 x 16 Ak 1.15	Total cfm cfm L/S Throw L/S	460 74/156 5/7	575 92/196 6/9	690 110/235 7.5/11	805 129/274 8.5/13	920 147/313 10/15	1035 166/352 11/16	1150 184/391 12/18	1265 202/430 14/20	1380 221/469 15/22
36 x 16 Ak 1.40	Total cfm cfm L/S Throw L/S	560 90/190 5.5/8	700 113/237 7/10	840 135/285 8.5/12	980 158/332 9.5/14	1120 180/380 11/16	1260 203/427 12/18	1400 225/475 14/20	1540 248/522 15/22	1680 270/570 17/24

Terminal velocity of 75 fpm.

TITUS® 300/350 Series Grilles & Registers

Louvered Grilles & Registers

Supply & Return Models

The new TITUS® 300/350 Series grilles and registers meet the demand for top quality design and construction, competitively priced.

This series is the TITUS standard grille and register line, built to today's requirements, in both all-steel and all-aluminum models.

When you specify TITUS 300/350 Series, you get these features:

- Blades are formed to a specially contoured cross-section. In supply grilles and registers, both ends of each blade are mounted in friction pivots which allow individual blade adjustment without loosening or rattling.
- Choice of 2 different blade spacings and 2 blade angles in return grilles and registers.
- Optional opposed-blade damper has a screwdriver adjustment accessible through the face of the register.
- Screw holes in the border are countersunk for smooth appearance after installation.

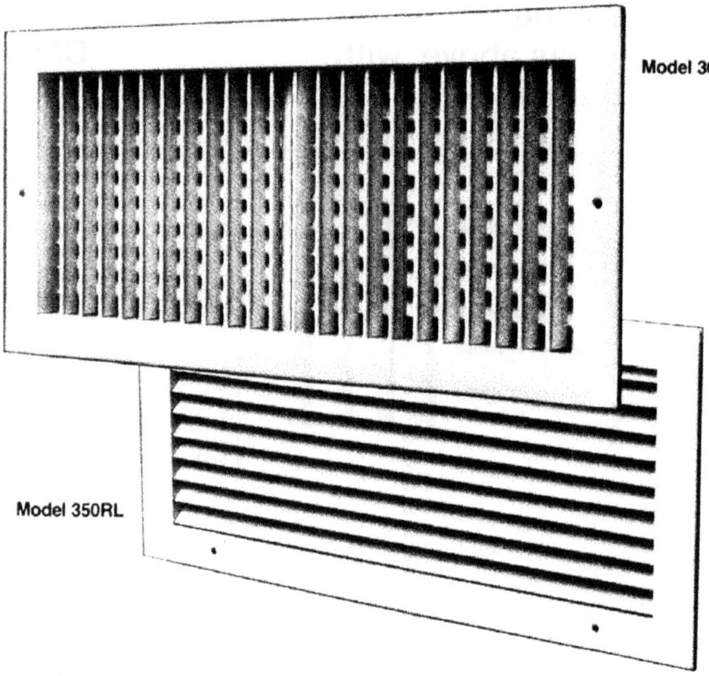

- Baked-on enamel finish similar to that on a household appliance. Metal surfaces are degreased, phosphatized and dried, then enameled by electrodeposition, followed by oven baking.

Standard Finish:

#25 White.

Additional 300/350 Information:

Supply Dimensions .... Page 5
Supply Performance 6-9
Return Dimensions 10, 11
Return Performance 12-15

300/350 Series Grilles & Registers • Available Models

Function	Description	All Steel		All Aluminum	
		Front Blades Horizontal	Front Blades Vertical	Front Blades Horizontal	Front Blades Vertical
Supply	Single Deflection • ¾″ Spacing	301RL	301RS	301FL	301FS
	Double Deflection • ¾″ Spacing	300RL	300RS	300FL	300FS
Return	35° Fixed Deflection • ¾″ Spacing	350RL	350RS	350FL	350FS
	0° Fixed Deflection • ¾″ Spacing	350ZRL	350ZRS	350ZFL	350ZFS
	35° Fixed Deflection • ½″ Spacing	355RL	355RS	355FL	355FS
	0° Fixed Deflection • ½″ Spacing	355ZRL	355ZRS	355ZFL	355ZFS

Available Sizes: 6″ × 4″ through 48″ × 36″ in 1″ increments.
Balancing Devices: Optional Model AG-15 opposed blade damper.
Optional Finish: #01 Aluminum. For aluminum grilles and registers only.

TITUS® 300/350 Series Grilles & Registers

odels 300R, 300F, 301R, 301F • Supply

NC-20 NC-30 NC-40 NC-50

Core, Area, Square Feet	Nominal Duct Size, Inches													
		Core Vel.		300	400	500	600	700	800	1000	1200	1400	1600	1800
		Vel. Press.		.006	.010	.016	.022	.030	.040	.062	.090	.122	.159	.202
		Total Press.	0°	.017	.029	.047	.065	.088	.117	.181	.265	.358	.467	.593
			22½°	.019	.032	.052	.073	.098	.132	.204	.297	.403	.523	.666
			45°	.027	.049	.080	.109	.149	.199	.307	.447	.605	.788	1.000
.12	6 × 4	CFM		35	50	60	70	85	95	120	140	165	190	210
		NC		—	—	—	12	18	22	28	34	39	43	46
		Throw, Feet	0°	4-6-12	5-8-14	7-10-16	8-12-17	9-13-19	11-14-20	13-16-22	14-17-24	15-19-26	16-20-28	17-22-30
			22½°	3-5-10	4-6-11	6-8-13	6-10-14	7-10-15	9-11-16	10-13-18	11-14-19	12-15-21	13-16-22	14-18-24
			45°	2-3-6	3-4-7	3-5-8	4-6-9	5-7-9	5-7-10	6-8-11	7-9-12	8-9-13	8-10-14	9-11-15
.18	8 × 4 6 × 6	CFM		55	70	90	110	125	145	180	215	250	290	325
		NC		—	—	—	15	20	24	30	36	41	45	48
		Throw, Feet	0°	4-7-13	6-8-15	7-11-17	9-13-19	10-15-20	11-16-22	14-17-24	15-19-26	17-21-29	18-22-31	19-24-33
			22½°	3-6-10	5-6-12	6-9-14	7-10-15	8-12-16	9-13-18	11-14-19	12-15-21	14-17-23	14-18-25	15-19-26
			45°	2-3-7	3-4-8	4-5-9	4-7-10	5-7-10	6-8-11	7-9-12	8-10-13	8-10-14	9-11-15	10-12-16
.22	10 × 4 7 × 6	CFM		65	90	110	130	155	175	220	265	310	350	395
		NC		—	—	—	15	20	24	30	36	41	45	48
		Throw, Feet	0°	4-7-14	7-10-17	8-12-19	9-15-21	11-16-23	13-17-24	16-19-27	17-21-29	19-23-32	20-25-34	21-26-36
			22½°	3-6-11	6-8-14	6-10-15	7-12-17	9-13-18	10-14-19	13-15-22	14-17-23	15-18-26	16-20-27	17-21-29
			45°	2-4-7	3-5-9	4-6-10	5-7-10	6-8-11	6-9-12	8-10-13	9-11-15	9-12-16	10-12-17	11-13-18
.26	12 × 4 8 × 6	CFM		80	105	130	155	180	210	260	310	365	415	470
		NC		—	—	—	16	21	25	31	37	42	46	49
		Throw, Feet	0°	5-8-16	7-11-19	9-13-21	10-16-23	12-17-24	14-19-26	17-21-29	19-23-32	20-25-35	22-26-37	23-27-40
			22½°	4-6-13	6-9-15	7-10-17	8-13-18	10-14-19	11-15-21	14-17-23	15-18-26	16-20-28	18-21-30	18-22-32
			45°	3-4-8	4-5-9	4-7-10	5-8-11	6-9-12	7-9-13	8-11-15	9-12-16	10-13-17	11-13-18	12-14-20
.34	16 × 4 10 × 6	CFM		100	135	170	205	240	270	340	410	475	545	610
		NC		—	—	—	16	21	25	31	37	42	46	49
		Throw, Feet	0°	5-9-18	8-12-21	10-15-24	12-19-26	14-20-28	16-22-30	20-24-33	22-26-37	23-28-40	25-30-42	26-32-45
			22½°	4-7-14	6-10-17	8-12-19	10-15-21	11-16-22	13-18-24	16-19-26	18-21-30	18-22-32	20-24-34	21-26-36
			45°	3-4-9	4-6-11	5-8-12	6-9-13	7-10-14	8-11-15	10-12-17	11-13-18	12-14-20	12-15-21	13-16-22
.39	18 × 4 12 × 6 8 × 8	CFM		115	155	195	235	275	310	390	470	545	625	700
		NC		—	—	12	17	22	26	32	38	43	47	50
		Throw, Feet	0°	6-9-19	9-13-23	11-16-25	13-19-28	15-22-30	17-23-32	21-26-36	23-27-40	25-30-42	27-33-45	28-35-48
			22½°	5-7-15	7-10-18	9-13-20	10-15-22	12-18-24	14-18-26	17-21-29	18-22-32	20-24-34	22-26-36	22-28-38
			45°	3-5-10	4-6-11	5-8-13	7-10-14	8-11-15	9-12-16	11-13-18	12-14-20	12-15-21	13-16-23	14-17-24

TITUS® 300/350 Series Grilles & Registers

Models 300R, 300F, 301R, 301F • Supply

NC-20 NC-30 NC-40 NC-50

Core, Area, Square Feet	Nominal Duct Size, Inches													
		Core Vel.		300	400	500	600	700	800	1000	1200	1400	1600	1800
		Vel. Press.		.006	.010	.016	.022	.030	.040	.062	.090	.122	.159	.202
		Total Press.	0°	.017	.029	.047	.065	.088	.117	.181	.265	.358	.467	.593
			22½°	.019	.032	.052	.073	.098	.132	.204	.297	.403	.523	.666
			45°	.027	.049	.080	.109	.149	.199	.307	.447	.605	.788	1.000
.46	20 × 4 14 × 6 10 × 8	CFM		140	185	230	275	320	370	460	550	645	735	830
		NC		—	—	12	18	23	27	35	39	41	48	51
		Throw, Feet	0°	7-10-22	9-14-25	12-17-27	14-22-30	16-23-32	19-25-35	23-27-39	25-31-43	27-33-46	29-35-49	31-38-52
			22½°	6-8-18	7-11-20	10-14-22	11-18-24	13-18-26	15-20-28	18-22-31	20-25-34	22-26-37	23-28-39	25-30-42
			45°	3-5-11	5-7-12	6-9-14	7-11-15	8-11-16	10-13-17	11-14-20	12-15-21	14-17-23	14-18-24	15-19-26
.52	24 × 4 16 × 6	CFM		155	210	260	310	365	415	520	625	730	830	935
		NC		—	—	13	19	24	28	36	40	42	49	52
		Throw, Feet	0°	7-11-23	10-15-26	13-19-29	15-22-32	18-25-35	20-26-37	24-30-41	27-33-45	29-35-49	31-38-52	32-40-55
			22½°	6-9-18	8-12-21	10-15-23	12-18-26	14-20-28	16-21-30	19-24-33	22-26-36	23-28-39	25-30-42	26-32-44
			45°	3-5-11	5-7-13	6-9-15	8-11-16	9-12-17	10-13-18	12-15-21	13-16-23	14-18-24	15-19-26	16-20-28
.60	28 × 4 18 × 6 12 × 8 10 × 10	CFM		180	240	300	360	420	480	600	720	840	960	1080
		NC		—	—	14	20	25	29	37	41	43	50	53
		Throw, Feet	0°	7-12-24	11-16-28	14-20-31	16-24-34	19-27-37	22-29-40	26-32-45	29-35-48	31-38-52	23-40-56	35-43-59
			22½°	6-10-19	9-13-22	11-16-25	13-19-27	15-22-30	18-23-32	21-26-36	23-28-38	25-30-42	26-32-45	28-34-47
			45°	4-6-12	5-8-14	7-10-16	8-12-17	10-13-19	11-14-20	13-16-22	14-17-24	15-19-26	16-20-28	17-21-29
.69	30 × 4 20 × 6 14 × 8 12 × 10	CFM		205	275	345	415	485	550	690	830	965	1100	1240
		NC		—	—	15	20	25	29	37	41	43	50	53
		Throw, Feet	0°	8-13-26	12-17-30	15-22-34	18-26-37	21-29-40	24-31-43	28-34-47	30-38-52	33-40-56	35-43-60	37-45-63
			22½°	6-10-21	10-14-24	12-18-27	14-21-30	17-23-32	19-25-34	22-27-38	24-30-42	26-32-45	28-34-48	30-36-50
			45°	4-6-13	6-9-15	7-11-17	9-13-18	10-14-20	12-15-21	14-17-24	15-19-26	16-20-28	18-22-30	19-23-31
.81	36 × 4 22 × 6 16 × 8 14 × 10	CFM		245	325	405	485	565	650	810	970	1130	1300	1460
		NC		—	—	16	21	26	30	38	42	44	51	54
		Throw, Feet	0°	8-14-28	13-19-33	16-23-37	19-28-40	23-31-43	26-33-46	30-37-51	33-41-56	36-44-60	38-46-64	40-49-68
			22½°	6-11-22	10-15-26	13-18-30	15-22-32	18-25-34	21-26-37	24-30-41	26-33-45	29-35-48	30-37-51	32-39-54
			45°	4-7-14	6-9-16	8-12-18	10-14-20	11-15-22	13-17-23	15-19-26	17-20-28	18-22-30	19-23-32	20-25-34
.90	40 × 4 26 × 6 18 × 8 16 × 10 12 × 12	CFM		270	360	450	540	630	720	900	1080	1260	1440	1620
		NC		—	—	16	21	26	30	38	42	44	51	54
		Throw, Feet	0°	9-15-30	14-20-34	17-25-39	21-30-42	24-33-45	27-35-48	32-39-55	35-43-59	37-46-63	40-49-68	42-52-72
			22½°	7-12-24	11-16-27	14-20-31	17-24-34	19-26-36	22-28-38	26-31-44	28-34-47	30-37-50	32-39-54	34-42-58
			45°	5-8-15	7-10-17	9-13-19	10-15-21	12-16-23	14-17-24	16-20-27	17-21-29	19-23-32	20-24-34	21-26-36
1.18	34 × 6 24 × 8 20 × 10 16 × 12 14 × 14	CFM		355	470	590	710	825	945	1180	1420	1650	1890	2120
		NC		—	—	17	22	27	31	39	43	45	52	55
		Throw, Feet	0°	10-17-34	15-23-40	19-28-44	23-35-48	27-38-52	31-40-56	36-45-62	40-48-67	43-52-73	45-56-78	48-59-83
			22½°	8-14-27	12-18-32	15-22-35	18-28-38	22-30-42	25-32-45	29-36-50	32-38-54	34-42-58	36-45-62	38-47-66
			45°	5-8-17	8-11-20	10-14-22	12-17-24	13-19-26	15-20-28	18-22-31	20-24-34	21-26-36	23-28-39	24-30-41

TITUS® 300/350 Series Grilles & Registers

odels 300R, 300F, 301R, 301F • Supply

NC-20 NC-30 NC-40 NC-50

Core, Area, Square Feet	Nominal Duct Size, Inches													
		Core Vel.	300	400	500	600	700	800	1000	1200	1400	1600	1800
		Vel. Press.	.006	.010	.016	.022	.030	.040	.062	.090	.122	.159	.202
		Total Press. 0°	.017	.029	.047	.065	.088	.117	.181	.265	.358	.467	.593
		Total Press. 22½°	.019	.032	.052	.073	.098	.132	.204	.297	.403	.523	.666
		Total Press. 45°	.027	.049	.080	.109	.149	.199	.307	.447	.605	.788	1.000
1.34	36 × 6 18 × 12 16 × 14	CFM	400	535	670	805	940	1070	1340	1610	1880	2140	2410
		NC	—	—	18	23	28	33	38	44	49	53	56
		Throw, Feet 0°	11-18-36	16-24-42	20-30-47	24-37-51	28-40-56	32-43-59	39-47-65	42-52-72	45-56-78	48-60-83	51-63-89
		Throw, Feet 22½°	9-14-29	13-19-34	16-24-38	19-30-41	22-32-45	26-34-47	31-38-52	34-42-58	36-45-62	38-48-66	41-50-71
		Throw, Feet 45°	6-9-18	8-12-21	10-15-23	12-18-25	14-20-28	16-21-29	19-23-33	21-26-36	23-28-39	24-30-42	26-32-44
1.60	30 × 8 24 × 10 22 × 12 18 × 14 16 × 16	CFM	480	640	800	960	1120	1280	1600	1920	2240	2560	2880
		NC	—	—	18	2	28	33	38	44	49	53	56
		Throw, Feet 0°	13-20-40	18-26-46	22-32-51	27-39-56	31-43-60	35-46-64	42-51-72	46-56-79	49-61-85	53-65-91	56-69-97
		Throw, Feet 22½°	10-16-32	14-21-37	18-26-41	22-31-45	25-34-48	28-37-51	34-41-58	37-45-63	39-49-68	42-52-73	45-55-78
		Throw, Feet 45°	6-10-20	9-13-23	11-16-25	13-20-28	15-22-30	17-23-32	21-26-36	23-28-39	25-30-43	26-32-46	28-35-48
2.08	40 × 8 36 × 10 30 × 12 24 × 14 20 × 16 18 × 18	CFM	625	830	1040	1250	1460	1660	2080	2500	2910	3330	3740
		NC	—	—	19	24	29	34	39	45	50	54	57
		Throw, Feet 0°	14-23-45	20-30-52	26-38-58	30-44-63	35-49-68	40-53-73	48-59-82	52-64-90	56-69-97	60-75-104	64-79-110
		Throw, Feet 22½°	11-18-36	16-24-42	21-30-46	24-35-50	28-39-54	32-42-58	38-47-66	42-51-72	45-55-78	48-60-83	51-63-88
		Throw, Feet 45°	7-11-23	10-15-26	13-19-29	15-22-32	17-25-34	20-36-37	24-29-41	26-32-45	28-35-48	30-37-52	32-40-55
2.45	48 × 8 32 × 12 26 × 14 24 × 16 20 × 18 20 × 20	CFM	735	980	1220	1470	1720	1960	2450	2940	3430	3920	4410
		NC	—	—	20	25	30	35	40	46	51	55	58
		Throw, Feet 0°	15-25-49	22-33-57	27-40-62	32-48-68	38-54-74	43-57-80	52-64-89	57-70-97	61-76-106	65-81-113	70-87-120
		Throw, Feet 22½°	12-20-39	18-26-46	22-32-50	26-38-54	30-43-59	34-46-64	42-51-71	46-56-78	49-61-85	52-65-90	56-70-96
		Throw, Feet 45°	7-12-24	11-16-28	14-20-31	16-24-34	19-27-37	22-28-40	26-32-45	28-35-49	32-38-53	33-42-56	35-43-60
3.11	48 × 10 40 × 12 36 × 14 30 × 16 26 × 18 24 × 20	CFM	935	1240	1560	1870	2180	2490	3110	3730	4350	4980	5600
		NC	—	14	21	26	31	35	41	47	52	56	59
		Throw, Feet 0°	17-27-55	24-36-63	34-45-71	41-53-78	47-60-84	48-64-90	58-2-100	64-79-110	69-86-118	74-92-128	79-97-135
		Throw, Feet 22½°	14-22-44	10-29-50	27-36-57	33-42-62	38-48-67	38-51-72	46-58-80	51-63-88	55-69-94	59-74-102	63-78-108
		Throw, Feet 45°	8-14-28	12-18-31	17-22-35	20-26-39	23-30-42	24-32-45	29-36-50	32-40-55	35-43-59	37-46-64	40-49-67
3.61	48 × 12 36 × 16 30 × 18 28 × 20 24 × 24	CFM	1080	1440	1800	2170	2530	2890	3610	4330	5050	5780	6500
		NC	—	15	22	27	32	36	42	48	53	57	60
		Throw, Feet 0°	18-29-59	26-38-68	32-47-76	38-56-84	44-65-90	51-69-97	63-78-108	69-86-118	75-93-128	80-99-137	86-105-146
		Throw, Feet 22½°	14-23-47	21-30-54	26-38-61	30-45-67	35-52-72	41-55-78	50-62-86	55-69-94	60-74-102	60-79-110	69-84-117
		Throw, Feet 45°	9-14-29	13-19-34	16-23-38	19-28-42	22-32-45	25-35-48	31-39-54	35-43-59	38-46-64	40-50-69	43-52-73
4.29	48 × 14 36 × 18 32 × 20 28 × 24	CFM	1290	1720	2140	2570	3000	3430	4290	5150	6010	6860	7720
		NC	—	16	23	28	33	37	43	49	54	58	61
		Throw, Feet 0°	19-31-64	28-41-74	35-50-83	42-60-91	49-71-98	56-76-106	69-85-118	76-93-130	82-102-140	88-108-149	92-115-158
		Throw, Feet 22½°	15-25-51	22-33-59	28-40-66	34-48-73	39-57-78	45-61-85	55-68-94	61-74-104	66-82-112	70-86-119	74-92-126
		Throw, Feet 45°	10-15-32	14-20-37	17-25-42	21-30-46	24-35-49	28-38-53	34-43-59	38-47-65	41-51-70	44-54-75	46-57-79

NC-60

TITUS® 300/350 Series Grilles & Registers

Models 300R, 300F, 301R, 301F • Supply

Core, Area, Square Feet	Nominal Duct Size, Inches			NC-20			NC-30		NC-40		NC-50		NC-60	
		Core Vel.		300	400	500	600	700	800	1000	1200	1400	1600	1800
		Vel. Press.		.006	.010	.016	.022	.030	.040	.062	.090	.122	.159	.202
		Total Press.	0°	.017	.029	.047	.065	.088	.117	.181	.265	.358	.467	.593
			22½°	.019	.032	.052	.073	.098	.132	.204	.297	.403	.523	.666
			45°	.027	.049	.080	.109	.149	.199	.307	.447	.605	.788	1.000
4.65	48 × 16 36 × 20 30 × 24	CFM		**1400**	**1860**	**2320**	**2790**	**3260**	**3720**	**4650**	**5580**	**6510**	**7440**	**8370**
		NC		—	16	23	28	33	37	43	49	54	58	61
		Throw, Feet	0°	20-**33**-67	29-**43**-78	36-**54**-87	44-**65**-95	51-**74**-103	58-**79**-110	72-**89**-123	79-**97**-135	86-**105**-146	91-**113**-156	96-**120**-164
			22½°	16-**26**-54	23-**34**-62	29-**43**-70	35-**52**-76	41-**59**-82	46-**63**-88	58-**71**-98	63-**78**-108	69-**84**-117	73-**90**-125	77-**96**-131
			45°	10-**16**-33	15-**22**-39	18-**27**-43	22-**32**-48	25-**37**-52	29-**40**-55	36-**44**-61	39-**49**-67	43-**52**-73	46-**56**-78	48-**60**-82
6.25	48 × 20 30 × 30	CFM		**1880**	**2500**	**3120**	**3750**	**4380**	**5000**	**6250**	**7500**	**8750**	**10,000**	**11,200**
		NC		—	18	25	30	35	39	45	51	56	60	63
		Throw, Feet	0°	23-**37**-78	33-**49**-90	42-**62**-100	50-**75**-103	58-**86**-119	67-**93**-128	84-**104**-143	92-**113**-156	98-**123**-169	106-**132**-180	112-**140**-192
			22½°	18-**30**-62	26-**39**-72	34-**50**-80	40-**60**-82	46-**69**-95	54-**74**-102	67-**83**-114	74-**90**-125	78-**98**-135	85-**105**-140	90-**112**-153
			45°	12-**19**-39	17-**25**-45	21-**31**-50	25-**37**-51	29-**43**-60	34-**46**-64	42-**52**-72	46-**57**-78	49-**61**-85	53-**66**-90	56-**70**-96

Correction Factors (Table at Right)

- Throw and total pressure corrections are multipliers. The NC correction is an addition.
- A_k is the flow factor. A_c is the core area from the main table.

Model	Damper	A_k/A_c	Throw	Total Pressure	NC
300R, 300F	With	.77	1.00	1.00	0
301R, 301F	Without	.82	.98	.88	−2

Performance Table Notes

- Data were obtained from tests conducted in accordance with ISO Standard 5219, ISO Standard 3741 and ADC Test 1062 GRD84.
- All pressures are in inches of water.
- Core velocities are in feet per minute.
- Throw values are given for terminal velocities of 150, 100 and 50 fpm.
- Each NC value represents the noise criterion curve which will not be exceeded by the sound pressure in **any** of the octave bands, 2nd through 7th. Each NC value is based on a room absorption of 10 dB, re 10^{-12} watts. Each NC value is further based on a single grille or register operating at a 0 degree deflection setting. For deflection settings of 22½ and 45 degrees, increase the stated sound levels by 1 and 7 NC respectively.
- Shaded dividing lines denote ranges of NC values.
- The stated deflection settings refer to the horizontal deflection (spread) of the air itself — not the blade position. For a 20 degree **upward** deflection, use the throw rating for the 0 degree setting and the total pressure for the 22½ degree horizontal setting.
- Dash (—) in space indicates NC value less than 10.

Variable Air Volume Applications

All TITUS® supply grilles and registers can be applied to variable air volume systems with excellent results. For detailed selection methods, consult your TITUS representative.

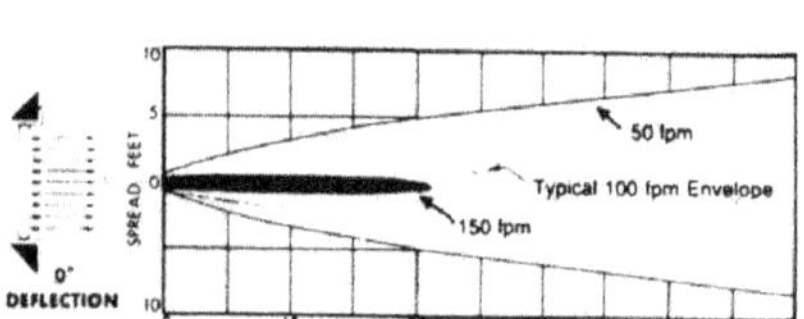

Spread versus throw at 0° horizontal deflection angle. (Plan view of air jet leaving wall mounted grille or register.)

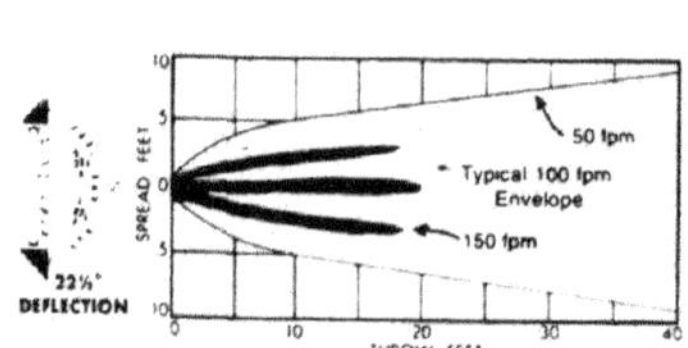

Spread versus throw at 22½° horizontal deflection angle. (Plan view of air jet leaving wall mounted grille or register.)

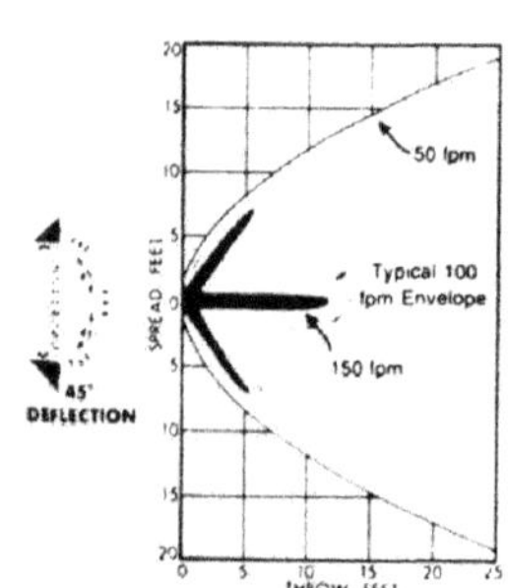

Spread versus throw at 45° horizontal deflection angle. (Plan view of air jet leaving wall mounted grille or register.)

TITUS® Balancing Devices • Introduction

ampers
qualizing Grids
urning Vanes
xtractors

fective air balancing begins th the balancing provisions at are designed into the 'AC system.

US® balancing devices are:

- Easy to select and specify. mprehensive dimensions and scriptions are available. Many items n be ordered as accessories on US diffusers.
- Designed for quick access and justment.
- Accurate, smooth and predictable in ir response to adjustment.
- Engineered for the least possible ise, turbulence and air resistance.

te: Additional TITUS balancing vices not shown in this catalog:

- Model ESM-1000 balancing station with velocity sensor and gauge taps.
- Model AG-606 pressure reducing valve with manual or pneumatic operator.

that balancing procedures can be rformed at the diffusers with curacy based on careful prior sting, published flow factors are ailable for all TITUS diffusers.

ease contact your TITUS presentative for complete details.

Model AG-100
Sliding Blade Damper

Model AG-85
Butterfly Damper

Model AG-75
Opposed Blade Damper

Model EG
Equalizing Grid

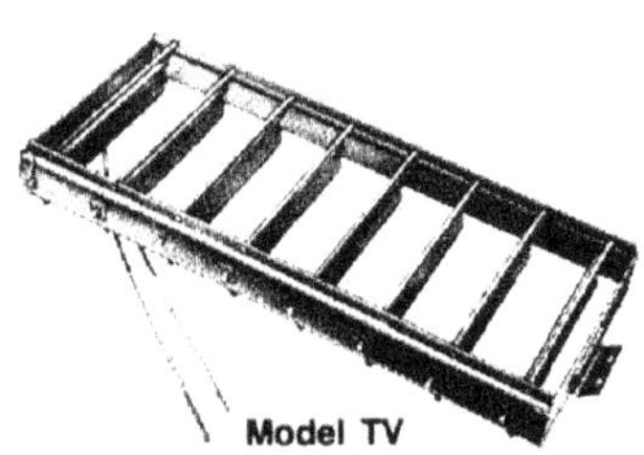

Model TV
Turning Vanes

Model AG-65
Damper with Grid

Model AG-95
Opposed Blade Damper

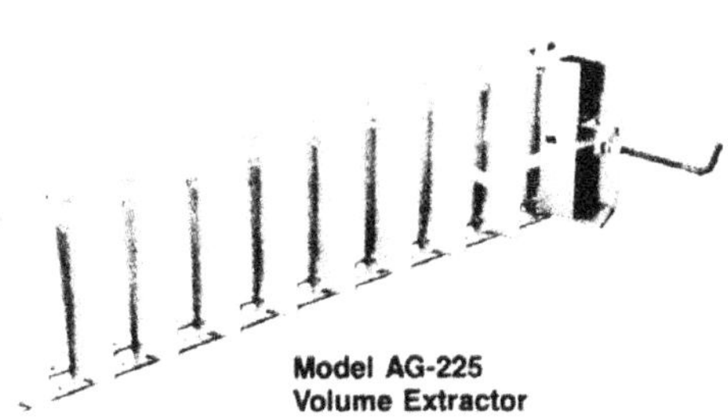

Model AG-225
Volume Extractor

TITUS® Exhaust & Return Grilles & Registers

Bar Return Grilles All Steel • Heavy Duty

⅜" Spacing • 0° Deflection

Model 30RL. One set of fixed blades parallel to **long** dimension. Optional model AG-35 (steel) or AG-35-AA (aluminum) opposed blade damper.

Used in gymnasiums and other locations where an unusually strong, impact resistant return grille is required.

Model 30RS. Same as 30RL except blades parallel to **short** dimension.

½" Spacing • 38° Deflection

Models 33RL and 33 RS. Same as 30RL and 30RS except with angled blades and wider spacing.

Standard Finish:

#25 White.

Extremely Heavy Construction:
16 Gauge Border, 14 Gauge Blades.

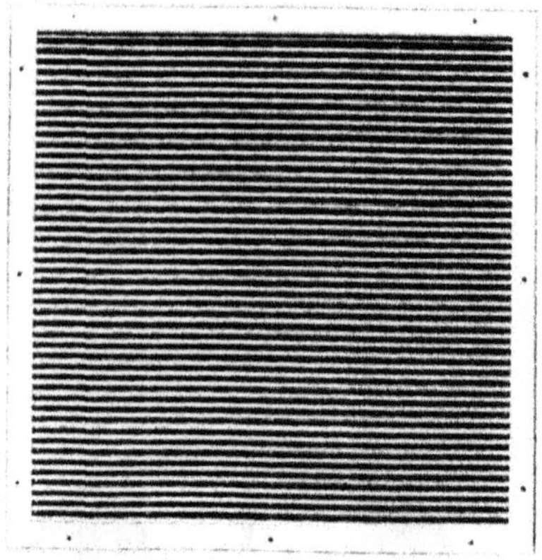

Face View, Model 33RS, 38° Deflection

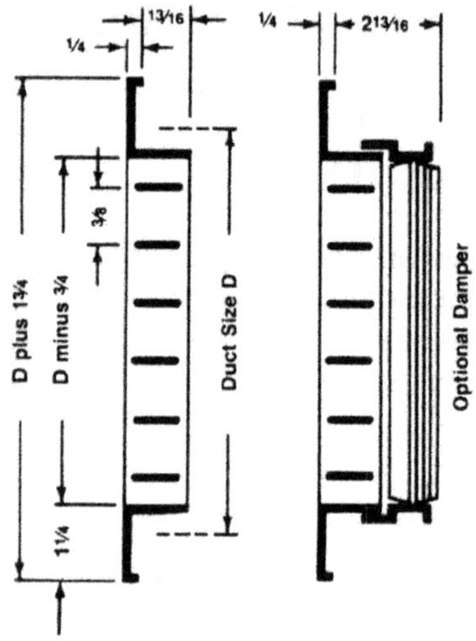

End View, Model 30RS, 0° Deflection

Door Return Grilles Sight Proof

All Steel

Model T-700. Offers the largest free area possible with a completely sight proof design. Widely used, not only in doors and partitions, but also in place of conventional exhaust and return grilles where it is important that the interior of the plenum or duct be concealed.

The inverted-V blades not only block vision, but also provide stiffness. Being mounted in the lower portion of a door, a door grille is naturally subjected to bumps and kicks. Here again, T-700 durability assures good looks and long service.

Heavy duty 20 gauge steel blades.

All Aluminum

Model CT-700. Similar to T-700 except all aluminum construction. Not available with channel type border.

Standard Finish:

#25 white.

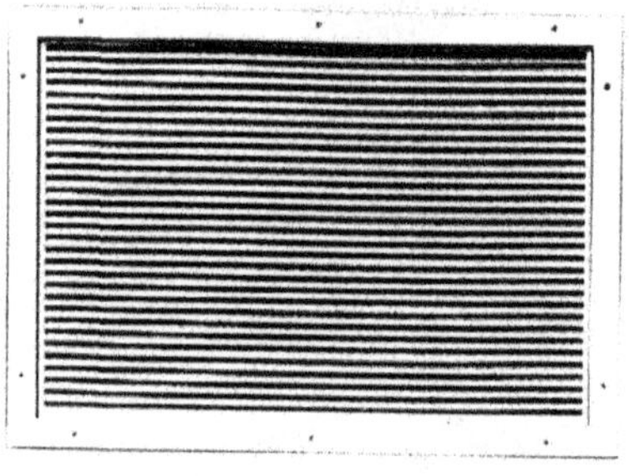

Model CT-700 with Flanged and Auxiliary Borders

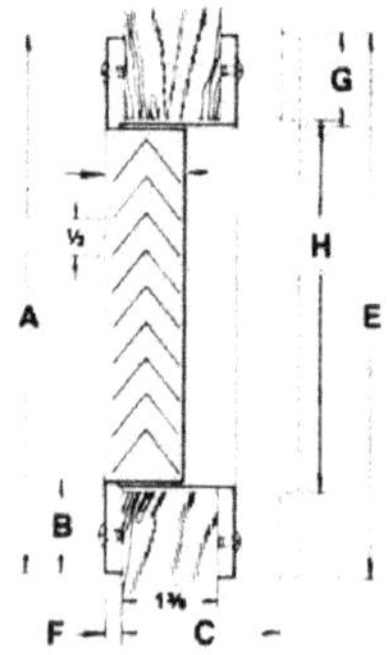

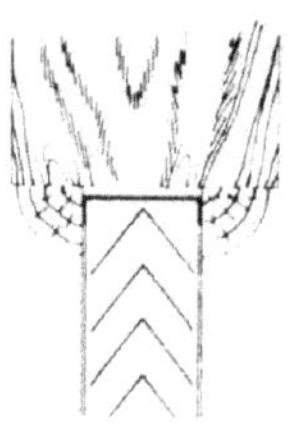

Channel Border

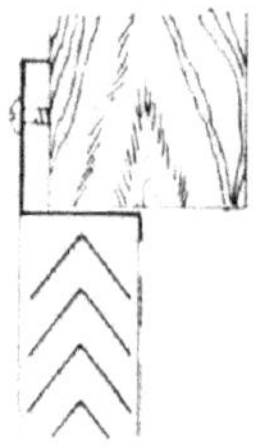

Flanged Border

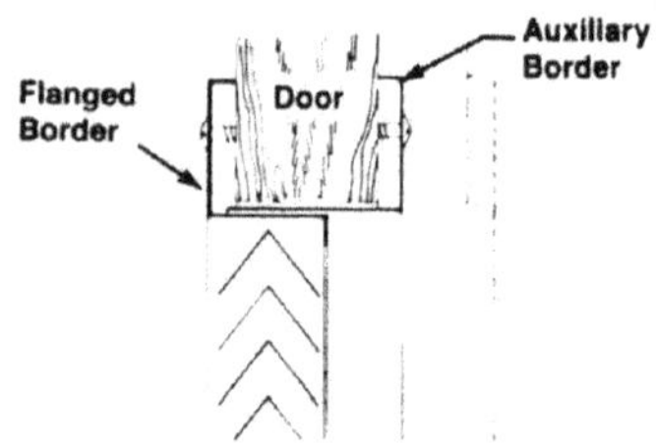

Flanged & Auxiliary Borders

Note: D = Listed Opening Size. H = Required Actual Opening Size.

Dimension	Model T-700 • Steel			Model CT-700 • Aluminum		
	Flange	Flange plus Aux.	Channel	Flange	Flange plus Aux.	Channel
A	D + 2⅛	D + 2⅛	—	D + 2¼	D + 2¼	NA
B	1¼	1¼	1¼	1⁷⁄₁₆	1⁷⁄₁₆	NA
C	—	2½	—	—	2¼	NA
D	Listed Opening Size			Listed Opening Size		NA
E	—	D + 2⅜	—	—	D + 2⁷⁄₁₆	NA
F	¼	¼	—	³⁄₁₆	³⁄₁₆	NA
G	—	—	—	—	1¼	NA
H	D − ¼	D + ⅛	D − ⅛	D − ⅛	D + ⅛	NA

NA symbol indicates channel border is not available in aluminum.

Dimensions are in inches.

TITUS® Exhaust & Return Grilles & Registers

del 30R • Bar • ⅜″ Spacing • 0° Deflection

NC-20 NC-30

ore rea, q. Ft.	Nominal Duct Size, Inches	Core Vel. / Vel. Press. / Neg. SP	400 / .010 / .029	500 / .016 / .046	600 / .022 / .064	700 / .031 / .089	800 / .040 / .116	900 / .050 / .145	1000 / .062 / .180	1100 / .075 / .218	1200 / .090 / .261	1400 / .122 / .354	
.15	7×4	CFM	60	75	90	105	120	135	150	165	180	210	
	6×5	NC	—	—	—	—	13	18	22	26	29	35	
.18	8×4 6×6	CFM	72	90	108	126	144	162	180	198	216	252	
	7×5	NC	—	—	—	10	15	19	23	27	31	37	
.22	10×4 7×6	CFM	88	110	132	154	176	198	220	242	264	308	
	8×5	NC	—	—	—	11	16	21	25	29	32	38	
.26	12×4 8×6	CFM	104	130	156	182	208	234	260	286	312	364	
	10×5	NC	—	—	—	12	17	22	26	30	33	39	NC-40
.34	16×4 10×6	CFM	136	170	204	238	272	306	340	374	408	476	
	12×5	NC	—	—	—	14	19	24	28	32	35	41	
.39	18×4 12×6	CFM	156	195	234	273	312	351	390	429	468	546	
	14×5 8×8	NC	—	—	—	15	20	25	29	33	36	42	
.46	20×4 14×6	CFM	184	230	276	322	368	414	460	506	552	644	
	16×5 10×8	NC	—	—	10	16	22	26	30	34	37	43	
.52	24×4 16×6	CFM	208	260	312	364	416	468	520	572	624	728	
	18×5	NC	—	—	11	17	22	27	31	35	38	44	
.60	28×4 18×6 10×10	CFM	240	300	360	420	480	540	600	660	720	840	
	20×5 12×8	NC	—	—	12	18	24	28	32	36	39	45	
.69	30×4 20×6 12×10	CFM	276	345	414	483	552	621	690	759	828	966	
	24×5 14×8	NC	—	—	13	19	25	29	33	37	40	46	
.81	36×4 22×6 14×10	CFM	324	405	486	567	648	729	810	891	972	1134	
	28×5 16×8	NC	—	—	14	21	26	30	34	38	42	48	
.90	40×4 26×6 16×10	CFM	360	450	540	630	720	810	900	990	1080	1260	
	30×5 18×8 12×12	NC	—	—	15	21	27	31	35	39	42	48	
1.07	48×4 30×6 14×12	CFM	428	535	642	749	856	963	1070	1177	1284	1498	
	36×5 18×10	NC	—	—	17	23	28	32	36	40	44	50	
1.18	34×6 20×10 14×14	CFM	472	590	708	826	944	1062	1180	1298	1416	1652	
	24×8 16×12	NC	—	10	17	23	28	33	37	41	44	50	NC-50
1.34	60×4 36×6 16×14	CFM	536	670	804	938	1072	1206	1340	1474	1608	1876	
	48×5 18×12	NC	—	11	18	24	29	34	38	42	45	51	
1.60	72×4 24×10 18×14	CFM	640	800	960	1120	1280	1440	1600	1760	1920	2240	
	30×8 22×12 16×16	NC	—	12	19	25	31	35	39	43	47	53	
2.08	72×5 40×8 30×12 20×16	CFM	832	1040	1248	1456	1664	1872	2080	2288	2496	2912	
	60×6 36×10 24×14 18×18	NC	—	14	21	27	33	37	41	45	48	54	
2.45	72×6 32×12 24×16	CFM	980	1225	1470	1715	1960	2205	2450	2695	2940	3430	
	48×8 26×14 20×18	NC	—	15	23	29	34	38	43	46	50	56	
2.78	36×12 26×16 22×20	CFM	1112	1390	1668	1946	2224	2502	2780	3058	3336	3892	
	30×14 24×18	NC	—	16	23	29	35	39	43	47	51	57	
3.11	60×8 40×12 30×16 24×20	CFM	1244	1555	1866	2177	2488	2799	3110	3421	3732	4354	
	48×10 36×14 26×18	NC	—	17	24	30	36	40	44	48	51	57	
3.61	72×8 48×12 30×18	CFM	1444	1805	2166	2527	2888	3249	3610	3971	4332	5054	
	60×10 36×16 24×24	NC	10	18	25	31	37	41	45	49	52	58	
4.29	48×14 32×20	CFM	1716	2145	2574	3003	3432	3861	4290	4719	5148	6006	
	36×18 28×24	NC	11	20	27	33	38	42	47	50	54	60	NC-60
5.58	72×12 48×18	CFM	2232	2790	3348	3906	4464	5022	5580	6138	6696	7812	
	60×14 36×24	NC	13	21	29	35	40	44	49	52	56	62	
6.25	72×14 48×20	CFM	2500	3125	3750	4375	5000	5625	6250	6875	7500	8750	
	60×16 30×30	NC	14	22	29	35	41	45	49	53	56	62	

ore velocities are in feet per minute.
ll pressures are in inches of water.
leg. SP is negative static pressure.
IC values are based on a room
sortion of 10 dB, re 10^{-12} watts.

- Shaded dividing lines denote ranges of NC values.
- Dash (—) in space indicates NC value less than 10.
- Data were obtained from tests conducted in accordance with ISO Standard 5219, ISO Standard 3741 and ADC Test Code 1062 GRD84.

Appendix 2

Summary of Tables Charts and Equations

A2-1 Velocity/Flow-Rate Equation

The air flow rate, the cross sectional area (grille opening or duct section) and the velocity are related by the following formula. (Area and velocity are readily measured and the flow rate is calculated.)

$$Q = A \times V$$

- Q is the CFM
- A is the cross-sectional area in square feet .
- V is the air velocity in feet per minute

A2-2 Approximate Air Terminal CFM

The approximate supply CFM that is required to cool or heat a room can be estimated by using the "sensible heat" equation.

$$CFM = \frac{\text{Room Sensible Cooling Load (Btuh)}}{1.1 \times (\text{Room db} - \text{Supply Air db})}$$

$$CFM = \frac{\text{Room Heating Load (Btuh)}}{1.1 \times (\text{Supply Air db} - \text{Room db})}$$

A2-3 Acceptable Face Velocity & NC Values

The table below provides a listing of acceptable face velocity and NC values for various applications.

Application	NC Level	Maximum Velocity
Private Residence		
Rural/Suburban	20 to 25	700
Urban	25 to 30	700
Apartments	30 to 35	700
Hotels/Motels		
Private room or suite	30 to 35	700
Meeting/Banquet room	30 to 35	700
Halls, Corredors, Lobbies	35 to 40	800
Service and support areas	40 to 45	900
Offices		
Executive	25 to 30	700
Conference rooms	25 to 30	700
Private	30 to 35	700
Open bay areas	35 to 40	800
Computer room	35 to 40	900
Business machine room	40 to 45	900
Waiting rooms	40 to 45	900
Retail Stores		
Clothing department	35 to 40	800
General sales areas	40 to 50	900
Supermarkets	40 to 50	900
Hospitals and Clinics		
Private/Semi-private rooms	25 to 30	700
Operating Rooms	25 to 30	700
Wards	30 to 35	700
Churches and Schools		
Sanctuaries	20 to 30	700
Libraries	30 to 35	700
Classrooms	30 to 35	700

Application	NC Level	Maximum Velocity
Churches and Schools (cont)		
Laboratories	35 to 40	800
Gyms and rec. areas	40 to 50	900
Kitchens	40 to 50	900
Public Assembly		
Concert halls	20 to 25	700
Legitimate theaters	20 to 25	700
Auditoriums	25 to 30	700
Motion picture theatres	30 to 35	700
Libraries/Museums/Courts	30 to 35	700
Coliseums	30 to 35	700
Bank/Post office	35 to 40	800
Lobbies	35 to 40	800
Swimming pools	40 to 55	1000
Transportation Terminals		
Ticketing areas	30 to 35	700
General Waiting	35 to 45	800
Studios		
Recording studios	20 to 25	700
Broadcasting studios	20 to 25	700
TV audience studios	30 to 35	700
Restaurants/Cafeterias/Lounges		
Restaurant eating area	35 to 40	800
Lounges	35 to 45	800
Night clubs	35 to 45	800
Cafeterias	40 to 50	900
Industrial Production Areas	Background Noise Sets Limits	

A2-4 Outlet Air Loading

The maximum CFM that can be delivered by a single outlet is usually determined by noise considerations (limits on face velocity, neck velocity or NC value) or throw considerations (maximum allowable throw) or both. Refer to the noise criteria table below and to the manufacturer's performance data to determine the maximum outlet air loading. Drop should also be considered. Refer to A2-5 and A2-6 below.

A2-5 Ceiling Outlets, Drop Considerations

Use the following table to determine how many ceiling outlets are required for any given supply CFM.

Maximum CFM Per Ceiling Outlet

Temp Diff	Ceiling Height (Ft)						
(F)	8	9	10	11	12	14	16
10	600	1500	4000				
15	425	1150	2250	4400			
20	250	700	1400	2800	4500		
25	175	500	975	1800	3000	6000	
30	100	200	550	1000	1500	3500	6500

A2-6 High Sidewall Outlets, Drop Considerations

Use the following generic drop tables to estimate the drop that is associated with high sidewall outlets. Note that the drop depends on the distance between the ceiling and the outlet and on the spread created by the vertical vane settings.

High Sidewall Near Ceiling
Horizontal Throw
0 To 20 Degree Spread

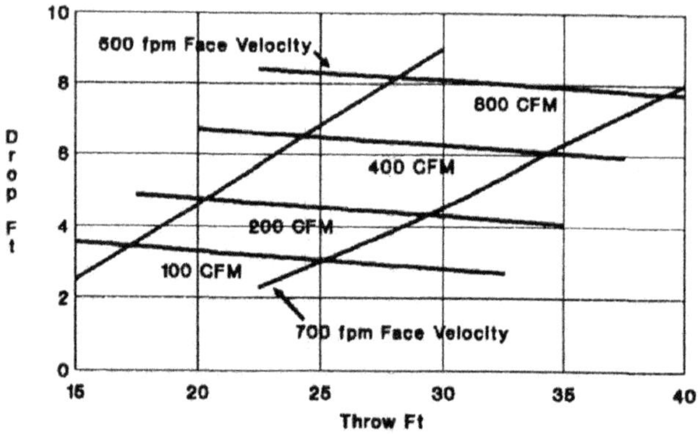

High Sidewall Near Ceiling
Horizontal Throw
30 To 45 Degree Spread

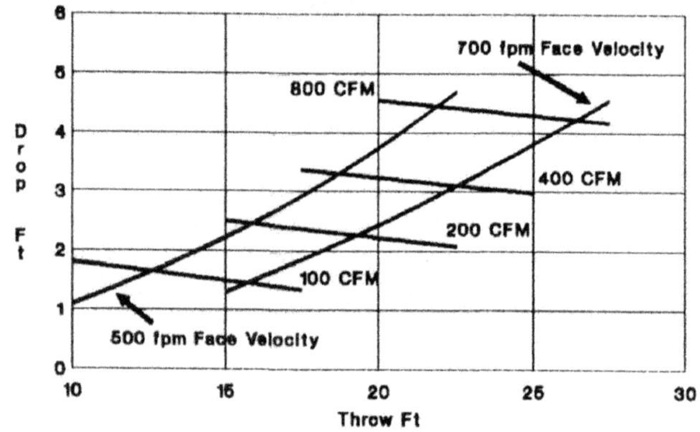

High Sidewall Below Ceiling
20 Degrees Up
Zero Degree Spread

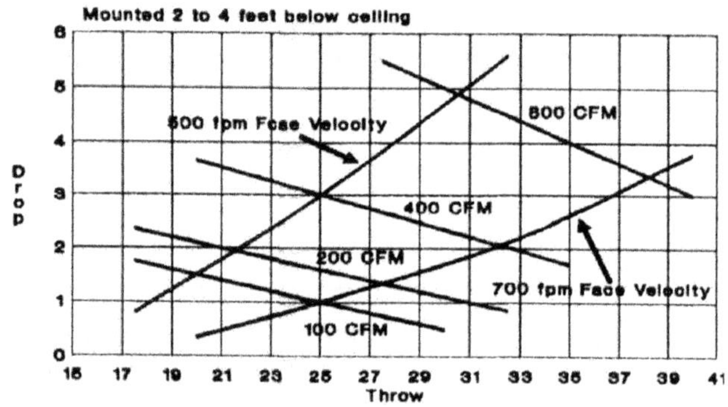

A2-7 High Sidewall, Spread vs Vane Setting

Use the following diagram to determine the spread that is associated with various vertical vane settings.

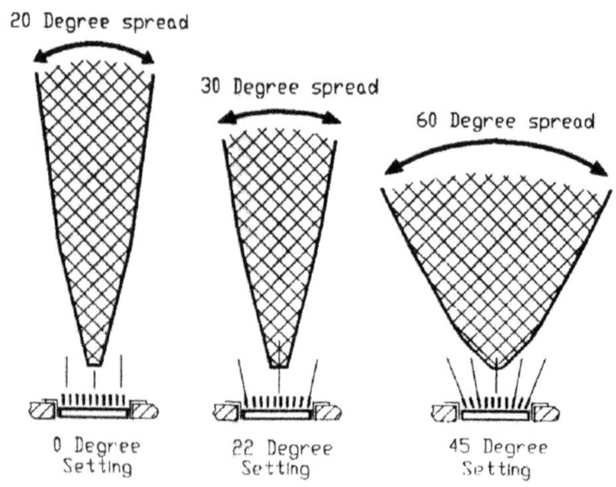

A2-8 Return Outlet Sizing

Return grille face velocity should be limited to 400 fpm. Filter grille and transfer grille face velocities should be limited to 300 fpm.

Index

A

Adjusted Dry Bulb Temperature 1-3

Air diffusion council (ADC)
  Test standard 10-2, 10-3

Air pattern
  Affect of vanes on ceiling outlet 4-4
  Affect of vanes on HSW 3-2, 3-3
  High sidewall outlet 3-2
    See also, Secondary air pattern
    See also, Total air pattern
    See also, Primary air pattern

Area factor 12-1

Attenuation
  Discussion 9-5
  Equipment noise 9-5, 9-6

B

Balance
  See, Test and balance 12-1

Baseboard outlet
  Horizontal discharge 5-2
  Location 10-1
  Selection 10-3
  Vertical discharge 5-1

Basement outlets 5-4

Beamed ceilings 4-5

C

Ceiling diffuser
  Horizontal discharge 4-2
  Location 10-1
  Selection 10-4, 10-5, 10-6
  Vertical discharge 4-2

Ceiling effect
  See, Surface effect

CFM
  See, Supply CFM

Comfort Chart 1-4

Comfort, effect of
  Building envelope 1-2
  Convection 1-1
  Neck and ankle temperature 1-3
  Radiation 1-1, 1-2
  Surrounding temperature 1-2
  Winter humidification 1-3
  Velocity 1-3

Convection 1-3

Core velocity 2-6

Cycles per second (CPS) 9-1

D

Decibels 9-1

Diffuser
  Baseboard outlet 2-2, 5-1
  Ceiling outlet 2-1, 4-1
  Floor outlet 2-2, 5-1
  Low sidewall outlet 5-1
  Location 10-1
  Supply CFM 10-1

Door grilles 7-1, 11-1

Drafts
  Caused by glass doors 5-5
  Caused by windows 5-6
  Definition 1-3

Drapes
  In front of glass doors 5-6
  In front of windows 5-6

Drop
  Ceiling diffuser 4-4
  Definition 2-2
  Drop charts 10-6, 10-7
  Effect of supply temperature
  Examples problems ...
    See, Section 10
  High sidewall outlet 3-3, 3-4

E

Entrance conditions
  Affect on noise 9-4
  Ceiling diffuser 4-1, 4-2
  Floor, baseboard outlets 5-2, 6-3, 6-4
  High sidewall outlet 3-1
  Returns 8-1, 8-2
  Sidewall outlets 6-3, 6-4

Entrained air 2-4

F

Face velocity
  Effect on noise 9-3
  Definition 2-6
  Example problems ...
    See, Section 10
  Maximum allowable 9-3

Filter grilles 7-1, 11-2

Floor outlet
Horizontal discharge 5-2
Location 10-1
Selection 10-3
Vertical discharge 5-1

Flow rate equation 2-6

Free area 2-5

Frequency 9-1

Frequency spectrum 9-1, 9-2

G

Glass doors 5-5, 5-6

Globe Thermometer 1-3

Grille
Definition 2-1
Location 10-1
Supply CFM 10-1

Gross area 2-5

H

Hertz 9-1

High sidewall outlet
Discussion 3-1
Selection 10-6, 10-7, 10-8

I

Inches of water 6-1

L

Loudness chart 9-2

Low sidewall outlet
Definition 5-1
Location 10-1

M

Manometer 6-1

Manufacturers data
Discussion 10-2
Test code 10-3
See also, Appendix 1

Mean Radiant Temperature 1-2

Metabolic Rate 1-1

Metabolism 1-2

N

Neck velocity 9-4, 10-4

Noise criteria (NC)
Acceptable levels 9-3, 10-2
Definition 9-2
Published ratings 9-2

Noise generation
Effect of dampers 9-4
Effect of entrance 9-4
Discussion 9-5
From equipment 9-5

O

Occupied zone 1-4

Octave band analyzer 9-2

P

Pressure drop
Accessories 6-3
Duct 6-2, 6-3
Effect of fitting geometry 6-3, 6-4
Example problems ...
See, Section 10
Loss coefficient 8-1
Outlets 6-2, 6-3
Returns 8-1

Primary air 2-2

Primary air pattern
Baseboard outlet 5-1, 5-2
Ceiling diffuser 4-1, 4-2
Floor outlet 5-1, 5-2
General discussion 2-4
High sidewall outlet 3-1
Low sidewall outlet 5-1, 5-2

R

Register
Definition 2-1
Location 10-1
Supply CFM 10-1

Returns
Effect on air patterns 7-1, 7-2
Combination supply/return 7-1
Door grilles 7-1, 11-1
Filter grilles 7-1, 11-2
Guidelines for location 7-2
Inlet vane adjustment 8-2
Location 3-4, 4-6, 11-1
No sight, no sound 7-1
Pressure loss 8-1, 11-1
Selection examples 11-2
Sizing 11-1
Transfer grilles 11-1, 11-2
Types 7-1

Rise
Definition 2-3
See also, Drop

S

Secondary air 2-2

Secondary air pattern
Discussion 2-3
High sidewall outlet 3-2

Selection procedures
Inlets 11-1
Outlets 10-1

Smudging 4-5

Sound power 9-1

Sound pressure 9-1

Spread
Effect of vane setting 10-7
Definition 2-3
Examples problems ...
See, Section 10

Stagnate Zone
See, Stratified zone

Stratified zone
Baseboard outlets 5-5
Ceiling diffuser 4-4, 4-5
Floor outlets 5-4
General discussion 2-4
High sidewall outlets 3-2
Low sidewall outlets 5-5

Static
Static pressure 6-1
Static regain 6-2

Supply CFM
Equations 10-1
Maximum CFM per outlet 10-5

Surface effect
Ceiling diffuser 4-4
Discussion 2-3

T

Temperature difference (TD) effect
Ceiling diffuser 4-4, 4-5, 10-4
High sidewall outlet 3-3, 3-4, 10-6

Temperature gradient
Vertical 1-4, 2-5, 4-3, 5-3, 5-4, 5-5, 7-2
See also, Stagnate zone

Terminal velocity 2-2

Test and balance
Area factor 12-1
Procedures 12-1

Throw
Definition 2-2
Examples problems ...
See, Section 10

Total air pattern
Baseboard 5-2, 5-3
Ceiling diffuser 4-2, 4-3
Floor outlet 5-2, 5-3
General discussion 2-4
High sidewall outlet 3-1
Low sidewall outlet 5-2, 5-3

Total pressure 6-1

Transfer grilles 11-1, 11-2

V

Velocity pressure 6-1

Vertical discharge
Ceiling diffuser 4-2, 4-3, 4-5
Floor outlet 5-1, 5-2, 5-3, 5-4, 5-5

W

Wave length 9-1